Kelley Wing

Algebra

Grades 5–9

Credits
Editors: Amy R. Gamble, Elise Craver
Copy Editor: Beatrice Allen

Visit *carsondellosa.com* for correlations to Common Core, state, national, and Canadian provincial standards.

PO Box 35665
Greensboro, NC 27425 USA
carsondellosa.com

ISBN 978-1-4838-0505-4
01-059141151

Table of Contents

Introduction

Competency in math skills creates a foundation for the successful use of math principles in the real world. Practicing algebra skills is the best way to improve at them.

This book was developed to help students practice and master a variety of algebraic concepts. The practice pages can be used first to assess proficiency and later as basic skill practice. The extra practice will help students advance to more challenging math work with confidence. Help students catch up, stay up, and move ahead.

Common Core State Standards (CCSS) Alignment

This book supports standards-based instruction and is aligned to the CCSS. The standards are listed at the top of each page for easy reference. To help you meet instructional, remediation, and individualization goals, consult the Common Core State Standards alignment chart on page 4.

Leveled Activities

Instructional levels in this book vary. Each area of the book offers multilevel math activities so that learning can progress naturally. There are three levels, signified by one, two, or three dots at the bottom of the page:

- Level I: These activities will offer the most support.
- Level II: Some supportive measures are built in.
- Level III: Students will understand the concepts and be able to work independently.

All children learn at their own rate. Use your own judgment for introducing concepts to children when developmentally appropriate.

Hands-On Learning

Review is an important part of learning. It helps to ensure that skills are not only covered but are internalized. The flash cards at the back of this book will offer endless opportunities for review. Use them for basic skill and enrichment activities to reinforce basic algebraic concepts.

There is also a certificate template at the back of this book for use as students excel at daily assignments or when they finish a unit.

Common Core State Standards Alignment Chart

Common Core State Standards*		Practice Page(s)
Operations and Algebraic Thinking		
Write and interpret numerical expressions.	5.OA.1–5.OA.2	17–19
Geometry		
Graph points on the coordinate plane to solve real-world and mathematical problems.	5.G.1	83, 84
Ratios and Proportional Relationships		
Understand ratio concepts and use ratio reasoning to solve problems.	6.RP.3	81, 82
Analyze proportional relationships and use them to solve real-world and mathematical problems.	7.RP.1, 7.RP.3	80–82
The Number System		
Apply and extend previous understandings of numbers to the system of rational numbers.	6.NS.5–6.NS.7	20–22, 83–85
Apply and extend previous understandings of operations with fractions to add, subtract, multiply, and divide rational numbers.	7.NS.1–7.NS.2	5–16
Expressions and Equations		
Apply and extend previous understandings of arithmetic to algebraic expressions.	6.EE.1–6.EE.3	17–19, 23–25, 38–40, 56–58
Reason about and solve one-variable equations and inequalities.	6.EE.6–6.EE.8	41–43, 85
Use properties of operations to generate equivalent expressions.	7.EE.1	53–67
Solve real-life and mathematical problems using numerical and algebraic expressions and equations.	7.EE.3, 7.EE.4	38–40, 71–73
Work with radicals and integer exponents.	8.EE.1	44–52
Understand the connections between proportional relationships, lines, and linear equations.	8.EE.6	86–91
Analyze and solve linear equations and pairs of simultaneous linear equations.	8.EE.7, 8.EE.8	26–40, 95–100
Functions		
Define, evaluate, and compare functions.	8.F.1–8.F.3	85–91
Number and Quantity: The Real Number System		
Extend the properties of exponents to rational exponents.	HSN-RN.2	101–103
Algebra: Seeing Structure in Expressions		
Interpret the structure of expressions.	HSA-SSE.2	59–67
Write expressions in equivalent forms to solve problems.	HSA-SSE.3	68–73
Algebra: Arithmetic with Polynomials and Rational Expressions		
Perform arithmetic operations on polynomials.	HSA-APR.1	53–61
Use polynomial identities to solve problems.	HSA-APR.4	59–61
Rewrite rational expressions.	HSA-APR.6, HSA-APR.7	74–79
Algebra: Reasoning with Equations and Inequalities		
Solve equations and inequalities in one variable.	HSA-REI.3, HSA-REI.4	26–37, 41–43, 68–73
Solve systems of equations.	HSA-REI.6	95–100
Represent and solve equations and inequalities graphically.	HSA-REI.10–HSA-REI.12	85, 92–97

Name ______________________________ 7.NS.A.1d

Adding Real Numbers

A number line is a great way to model the addition of real numbers.

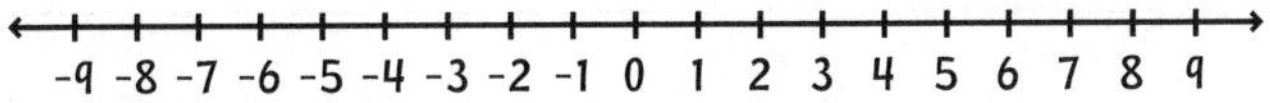

Add $3 + 5$. (Start at 3 and move 5 places to the right since 5 is positive.) The answer is 8.
Add $3 + ({}^{-}5)$. (Start at 3 and move 5 places to the left since 5 is negative.) The answer is ${}^{-}2$.
When adding a number that is positive, move to the right. When adding a number that is negative, move to the left.

Explain how to add the following numbers using a number line, and tell the result.

1. $7 + 8$ ______________________________

2. ${}^{-}5 + 9$ ______________________________

3. $10 + ({}^{-}7)$ ______________________________

4. ${}^{-}11 + ({}^{-}6)$ ______________________________

Use the number line to add the numbers.

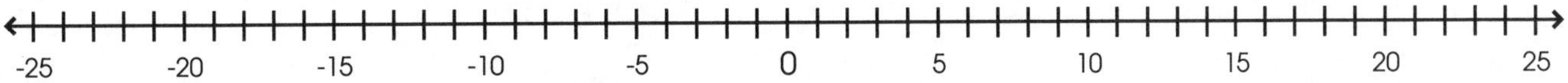

5. $4 + 9$

6. ${}^{-}5 + 13 + ({}^{-}11)$

7. ${}^{-}7 + ({}^{-}6)$

8. $14 + 10 + ({}^{-}3)$

9. $0 + ({}^{-}5)$

10. ${}^{-}6 + ({}^{-}9) + 10$

11. ${}^{-}3 + 3$

12. $8 + ({}^{-}2) + ({}^{-}4) + 6$

13. $7 + ({}^{-}12)$

14. ${}^{-}11 + 5 + ({}^{-}13) + 8$

15. ${}^{-}11 + 8 + ({}^{-}1)$

16. $12 + ({}^{-}6) + ({}^{-}14) + 17$

Name ____________________________________ 7.NS.A.1d

Adding Real Numbers

$^{-}7 + 6 = {}^{-}1$

Add.

1. $^{-}55 + (^{-}8) + (^{-}4) + 54 =$

2. $1 + (^{-}5) + (^{-}5) + 1 =$

3. $2.7 + (^{-}4.8) =$

4. $3.54 + 4.27 + 7.43 =$

5. $10 + 7 + (^{-}7) + (^{-}10) =$

6. $16 + 21 + (^{-}3) + 7 =$

7. $10 + 7 + (^{-}16) + 9 + (^{-}30) =$

8. $5.8 + 8.4 =$

9. $2.76 + (^{-}6.56) + (^{-}9.72) =$

10. $8 + (^{-}7) =$

11. $2\frac{3}{5} + 4\ 3\frac{3}{7} =$

12. $-8\frac{3}{5} + 3\frac{3}{7} =$

13. $2 + 5 + (^{-}3) =$

14. $-5\frac{3}{4} + (-2\frac{3}{4}) + 8 =$

15. $7.3 + (3.9) =$

16. $^{-}21 + 12 + (^{-}1) + (^{-}17) =$

17. $7.867 + (^{-}5.329) =$

18. $-2\frac{3}{5} + (-5\frac{3}{7}) + 3 =$

19. $3 + 12 + (^{-}13) + 36 =$

20. $-3\frac{1}{6} + (-9\frac{3}{12}) + 6 =$

Name ______________________ 7.NS.A.1d

Adding Real Numbers

Add.

1. $2\frac{3}{5} + (^-3\frac{2}{5}) + {}^-6 =$

2. $21 + 9 + (^-6) + 7 =$

3. $12 + (^-9) + 17 =$

4. $2.54 + {}^-5.87 + {}^-32.65 =$

5. $1.45 + 2.65 + (^-9.43) =$

6. $21 + 3 + (^-13) + 22 =$

7. $3 + (^-3) + 4 + (^-5) =$

8. $3.3 + (^-3.4) + 5.5 =$

9. $3.6 + (^-2.5) + {}^-5.5 =$

10. ${}^-0.6 + (^-0.56) + 3 =$

11. $3\frac{5}{8} + (^-1\frac{3}{4}) + 2 =$

12. $4.524 + 7.342 =$

13. $-7\frac{2}{4} + 2\frac{3}{4} =$

14. $34 + (^-13) + 18 + 0 + 34 =$

15. $8.43 + (^-10.98) + (^-3.23) =$

16. $2.54 + (^-5.21) + (^-6.34) =$

17. $-2\frac{1}{3} + (^-5\frac{7}{10}) + (^-7) =$

18. $-1\frac{2}{3} + (^-3\frac{3}{5}) + 4 =$

19. $2\frac{1}{2} + 6\frac{1}{2} =$

20. $4\frac{3}{5} + (^-3\frac{2}{5}) + (^-8) =$

Name ____________________ 7.NS.A.1c, 7.NS.A.1d

Subtracting Real Numbers

For all real numbers *a* and *b*, $a - b = a + (^-b)$. Simply stated: To subtract a number, add its opposite.

$7 - 10 = 7 + (^-10)$ $= {}^-3$	$5 - 8 + 3 - 1 = 5 + (^-8) + 3 + (^-1)$ $= 5 + 3 + (^-8) + (^-1)$ (group the numbers with like signs) $= 8 + (^-9)$ $= {}^-1$	$^-4 - (^-12)$ $= {}^-4 + 12$ $= 8$

Change each problem into an addition problem.

1. $7 - 9$
2. $^-6 - (^-4)$
3. $^-11 - 5$
4. $12 - (^-15)$
5. $8 - 3$
6. $22 - 5$
7. $4 - 11$
8. $^-4 - (^-9)$

Subtract.

9. $9 - 11$
10. $0 - (^-12)$
11. $^-5 - 4$
12. $6 - (^-6)$
13. $^-1 - (^-1)$
14. $^-7 - 6$
15. $3 - (^-5)$
16. $17 - 23$

Name ____________________

7.NS.A.1c, 7.NS.A.1d

Subtracting Real Numbers

$10 - (^{-}4) = 10 + 4 = 14$

Subtract.

1. $9 - (^{-}32) =$
2. $^{-}99 - (^{-}42) =$
3. $4 - (^{-}8) =$
4. $0 - 21 =$
5. $45 - 301 =$
6. $^{-}19 - 8 =$
7. $^{-}43 - 6 =$
8. $9 - (^{-}2) - 8 - 7 =$
9. $35 - 67 - 85 - 21 - 12 =$
10. $12 - 7 - (^{-}16) - 9 - (^{-}34) =$
11. $18 - (^{-}13) =$
12. $121 - 45 =$
13. $-\frac{4}{7} - \frac{1}{3} - (\frac{2}{3}) =$
14. $3.434 - 7.294 =$
15. $8 - 2.8 =$
16. $8 - (^{-}14) =$
17. $3.9 - 4.9 =$
18. $^{-}7 - (^{-}3) =$
19. $2.19 - 7.8 - 8.31 =$
20. $38 - 39 - (^{-}13) =$

Name ______________________________ 7.NS.A.1c, 7.NS.A.1d

Subtracting Real Numbers

Subtract.

1. $^{-}9 - (^{-}5) =$
2. $321 - (^{-}34) =$
3. $\frac{2}{3} - \frac{4}{5} =$
4. $\frac{3}{5} - \frac{7}{8} =$
5. $5.34 - 9.9 - 3.65 =$
6. $9.432 + 4.348 - 44.938 =$
7. $245 - 32 - (^{-}36) =$
8. $44 - 35 - 34 - 32 =$
9. $8 - (^{-}5) - 7 - 9 =$
10. $43 - 88 - 35 - 21 =$
11. $-\frac{2}{5} - \frac{3}{4} - (-\frac{4}{5}) =$
12. $^{-}45 - 5 =$
13. $-\frac{2}{3} - \frac{1}{3} - (-\frac{1}{3}) =$
14. $-\frac{4}{5} - \frac{1}{2} - \frac{2}{5} =$
15. $4 - 12.9 =$
16. $7 - (^{-}33) =$
17. $3.4 - 7.4 =$
18. $2.456 - 4.345 - 5.457 =$
19. $23 - (^{-}21) =$
20. $4.346 - 0.4537 =$

Name ______________________________ 7.NS.A.2a, 7.NS.A.2c

Multiplying Real Numbers

The **property of zero for multiplication** states that for all real numbers *a*, $a \bullet 0 = 0$ and $0 \bullet a = 0$. Simply stated, any real number multiplied by 0 is 0. For example, $0 \bullet 20 = 0$.

To multiply two real numbers with the same signs:

1. Multiply their absolute values.
2. The sign of their product is positive.

positive • positive = positive
(+) (+) (+)
$3 \bullet 12 = 36$

negative • negative = positive
(−) (−) (+)
$^-7 \bullet {}^-8 = 56$

To multiply two real numbers with different signs:

1. Multiply their absolute values.
2. The sign of their product is negative.

negative • positive = negative
(−) (+) (−)
$^-2 \bullet 5 = {}^-10$

positive • negative = negative
(+) (−) (−)
$4 \bullet {}^-8 = {}^-32$

Write the sign of the product for each number.

1. $(^-10)4$
2. $8(^-1)$
3. $(^-2)(^-3)$
4. $(7)(5)(^-3)$
5. $5(6)$
6. $(^-2)(^-7)$
7. $(^-12)(^-4)(^-1)$
8. $(^-6)(4)(^-2)$

Multiply to find each product.

9. $4(6)(^-1)$
10. $(^-1)(^-4)(^-3)$
11. $(^-\frac{1}{2})(2)$
12. $(7)(^-3)(0)$
13. $(5)(3)$
14. $(^-\frac{1}{8})(^-16)(4)$
15. $(^-9)(^-4)$
16. $(^-7)(7)$

Name ____________________

Multiplying Real Numbers

$(^-2)(^-3) = 6$

Multiply.

1. $4 \bullet 9 =$

2. $^-4 \bullet 12 =$

3. $(-\frac{5}{9})(8.8) =$

4. $(^-3)(0) =$

5. $(^-3)(^-9) =$

6. $6(23) =$

7. $(12)(^-3)(4) =$

8. $(^-5)(^-5)(^-5) =$

9. $(5)(2)(^-1) =$

10. $(7)(^-9)(^-12) =$

11. $(-\frac{2}{3})(^-1.6) =$

12. $^-7(^-7) =$

13. $(54.2)(^-3.55) =$

14. $(2.22)(^-1.11) =$

15. $(7.44)(3.2)(4.3) =$

16. $(2.4)(^-1.4) =$

17. $(-\frac{3}{5})(\frac{3}{5}) =$

18. $(-\frac{4}{5})(2.2) =$

19. $^-8 \bullet 12 =$

20. $(0)(2)(^-213) =$

Name ______________________

7.NS.A.2a, 7.NS.A.2c

Multiplying Real Numbers

Substitute and multiply.

$x = {}^-3, y = {}^-5, z = 0$

1. xy
2. ^-3yz
3. xyz
4. $4x(^-2y)$
5. $^-3(10xy)$
6. $2y(5y)$

$x = {}^-2.4, y = 3.1, z = {}^-4.8$

7. $4xz$
8. $yz(^-z)$
9. $^-5(xyz)$
10. $^-x(^-6z)$
11. $(xy)(^-3x)$
12. $(3yz)(^-7x)$

$x = \frac{1}{2}, y = -\frac{2}{3}, z = \frac{3}{5}$

13. $8(xy)$
14. xyz
15. $^-12x(^-5yz)$
16. $(xy)(3xz)$
17. $(^-2z)(3y)(6x)$
18. $^-z(10xyz)$

Name ____________________ (7.NS.A.2b, 7.NS.A.2c)

Dividing Real Numbers

The **multiplicative inverse property** states that for each nonzero a, there is exactly one number $\frac{1}{a}$ such that: $a \bullet \frac{1}{a} = 1$ and $\frac{1}{a} \bullet a = 1$.

The number $\frac{1}{a}$ is called the reciprocal of a.

For example, the reciprocal of 9 is $\frac{1}{9}$ and the reciprocal of $\frac{1}{2}$ is 2. Simply flip the number to find its reciprocal.

To divide the number a by the number b, multiply a by the reciprocal of b.

$a \div b = a \bullet \frac{1}{b}$ (The result will be the quotient of a and b.)

$12 \div 4 = 12 \bullet \frac{1}{4}$
$= 3$

$4x \div \frac{1}{4} = 4x \bullet 4$
$= 16x$

$15 \div \frac{3}{2} = 15 \bullet \frac{2}{3}$
$= 10$

To divide two nonzero real numbers:

1. The quotient is positive if both numbers have the same sign.
2. The quotient is negative if both numbers have different signs.

$^{-}15 \div 5 = {^{-}15} \bullet \frac{1}{5}$
$= {^{-}3}$

$3 \div \frac{1}{4} = 3 \bullet 4$
$= 12$

$1 \div -\frac{4}{5} = 1 \bullet -\frac{5}{4}$
$= -\frac{5}{4}$

Write the reciprocal of each number. Write none if it does not exist.

1. 2
2. 1
3. 0
4. $-\frac{1}{9}$
5. $-\frac{1}{4}$
6. $\frac{4}{3}$
7. $^{-}8$
8. 10

Divide.

9. $\frac{9}{3}$
10. $\frac{^{-}28}{7}$
11. $^{-}15 \div 3$
12. $\frac{^{-}7}{8} \div \frac{1}{8}$
13. $36 \div (^{-}4)$
14. $\frac{^{-}8}{^{-}8}$
15. $44 \div (^{-}11)$
16. $0 \div \frac{4}{5}$

Name ________________________________ 7.NS.A.2b, 7.NS.A.2c

Dividing Real Numbers

$9 \div 4.5 = 2$

Divide.

1. $\frac{49}{7} =$

2. $90 \div 15 =$

3. $(^{-}12) \div (9.9) =$

4. $(-\frac{2}{3}) \div (^{-}18) =$

5. $^{-}42 \div 7 =$

6. $45 \div (^{-}8) =$

7. $\frac{^{-}36}{4} =$

8. $(-\frac{3}{5}) \div (\frac{3}{5}) =$

9. $^{-}72 \div (9) =$

10. $^{-}21 \div (^{-}9) =$

11. $\frac{102}{17} =$

12. $0 \div (^{-}8) =$

13. $\frac{95}{5} =$

14. $\frac{63}{^{-}9} =$

15. $(^{-}3.4) \div (^{-}9.99) =$

16. $^{-}50 \div [40 \div (^{-}20)] =$

17. $(-\frac{4}{6}) \div (36) =$

18. $56 \div (^{-}28 \div 7) =$

19. $(^{-}38 \div 19) \div (^{-}2) =$

20. $^{-}45 \div [^{-}20 \div (^{-}4)] =$

Name ______________________________ 7.NS.A.2b, 7.NS.A.2c

Dividing Real Numbers

Substitute and divide.

$x = {}^{-}24,\ y = 6,\ z = {}^{-}8$

1. $\frac{x}{y}$

2. $\frac{yz}{{}^{-}x}$

3. $\frac{{}^{-}z}{xy}$

4. $\frac{4z}{5y}$

5. $\frac{{}^{-}xy}{{}^{-}yz}$

6. $\frac{2y}{{}^{-}5x}$

$w = 2.8,\ x = 1.4,\ y = {}^{-}8.4,\ z = {}^{-}11.2$

7. $(w) \div ({}^{-}x)$

8. $(z) \div ({}^{-}y)$

9. $\frac{x}{({}^{-}w)}$

10. $({}^{-}z) \div ({}^{-}w)$

11. $\frac{{}^{-}3y}{2x}$

12. $(z) \div ({}^{-}x)$

$x = -\frac{1}{4},\ y = -\frac{1}{3},\ z = \frac{4}{5}$

13. $(z) \div (x)$

14. $\frac{x}{{}^{-}z}$

15. $\frac{({}^{-}2x)}{(y)}$

16. $(y) \div ({}^{-}xz)$

17. $({}^{-}z) \div (y) \div ({}^{-}x)$

18. $\frac{{}^{-}z}{xy}$

Name ____________________

5.OA.A.1, 6.EE.A.2c

Order of Operations

When solving an equation, be sure to follow the **order of operations**.

1. Parentheses
2. Exponents
3. Multiplication & Division
4. Addition & Subtraction

$$14 - 54 \div 6 = 14 - 9 = 5$$

Solve.

1. $3 \times 15 \div 5 =$

2. $35 \div 5 - 9 =$

3. $3 + 2 \times 4 =$

4. $5 \times 2 \times 8 =$

5. $6 - 40 \div 8 =$

6. $12 - 30 \div 6 =$

7. $32 \div 4 \times 3 =$

8. $8 + 3 \times 2 =$

9. $4 + 12 \div 2 =$

10. $9 + 20 \div 5 =$

11. $15 - 75 \div 5 =$

12. $9 - 3 + 6 =$

13. $2 \times 8 \div 4 =$

14. $3 + 3 - 2 =$

Name ________________________________ 5.OA.A.1, 6.EE.A.1, 6.EE.A.2c

Order of Operations

Remember to follow the **order of operations**.

1. Parentheses
2. Exponents
3. Multiplication & Division
4. Addition & Subtraction

$(3^3 + 6 \times 5) - 2 = (27 + 6 \times 5) - 2 = (27 + 30) - 2 = 55$

Solve.

1. $(3^2 + 2 \times 3) \div 5 =$

2. $5^2 - 4^2 + 2 =$

3. $(4 + 2)^2 =$

4. $(11 - 8)^3 =$

5. $2(7 + 2) =$

6. $(9 - 7)^3 - (4 + 3) =$

7. $(14 - 6)2 =$

8. $4 + 3(12 - 9) =$

9. $5^2 - 2^3 =$

10. $3 \times 8 - (3 \times 2 + 7) =$

11. $(5^2 - 3 \times 5) \div 2 =$

12. $7 + 2^2(5 + 2) =$

13. $3 + 7^2 =$

14. $(2^2 + 3)^2 - 4 =$

15. $6 + 7 \times 3 - 9 \times 2 =$

16. $(2 \times 3) + (21 \div 7) =$

17. $7^2 - 2(3 \times 3 + 5) =$

18. $3 + (6 \times 2) =$

Name ______________________ 5.OA.A.1, 6.EE.A.1, 6.EE.A.2c

Order of Operations

Solve.

1. $8 - 4 \bullet 5(3 - 2) + 3 =$

2. $12 \div (2 - 7) + 7 =$

3. $(14 - 9) + 4 =$

4. $\frac{3^2 - 5 \bullet 7 - 4^2}{(^-4 - 7 - 12) + 8} =$

5. $9(3 \div 3) + 4(^-5 \bullet 9) \div 3 =$

6. $3 - (6 \bullet 6) - 3 \bullet 0 =$

7. $36 \div 9 - 8 + 21 \div 3 =$

8. $5(3 - 8) \bullet 3 + 8 - 3 =$

9. $3 \bullet 5 + 9 \bullet 7 =$

10. $\frac{(5 - 9)^2 + 2}{(7 - 8)^2 \bullet 3^2} =$

11. $4^2 + 3^2 - 7^2 =$

12. $\frac{3^2 - 10}{4^2 - 12} =$

13. $8^2 - \frac{26}{(4 + 9)} + 4 =$

14. $\frac{5 \bullet 7 - (3 + 4)}{^-(2^2) - 2^2 + 3^2} =$

15. $\frac{4 + 2 \bullet 3 + 4 - 3}{2^2 \bullet 3^2 - 3} =$

16. $\frac{3 + 10 - 19 + 32}{3^2 - 1 + 2^2} =$

17. $12 \div [3 + (6 + 3)] =$

18. $3 \bullet (0 - 7) + 8 \div 2^2 =$

Name ______________________________ 6.NS.C.5, 6.NS.C.6a, 6.NS.C.7

Real-Number Operations with Absolute Value

Every real number has an opposite. Opposite numbers are the same distance from 0 on a number line and lie on the opposite sides of 0. The opposite of a positive number is a negative number. The opposite of a negative number is a positive number. The numbers 3 and $^{-}3$ are opposites. Find their places on the real number line to the right. The additive inverse of a number is the same as the opposite of a number.

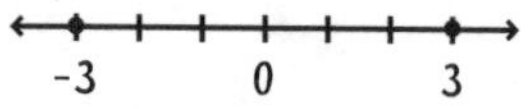

Remember, the opposite of 0 is simply 0 since it is neither positive nor negative.

The symbol $|x|$ is called the absolute value of x. The absolute value of a number is the distance between the number and 0 on a number line. The absolute value of a number, whether positive or negative, is always positive.

$|10| = 10$ $|^{-}2| = 2$ $|^{-}102| = 102$ $^{-}|12| = {}^{-}12$

Note: The answer to $^{-}|12|$ is $^{-}12$ because the absolute value of 12 is 12. But, it is multiplied by a negative, which resulted in $^{-}12$.

Write the opposite of each number.

1. 5
2. $^{-}2.6$
3. $^{-}40$
4. 2.8

Write a real number to represent each situation.

5. a gain of 12 yards
6. a temperature drop of 8°
7. a deposit of $89.26
8. a withdrawal of $75

Write the absolute value of each number.

9. $|10|$
10. $|23|$
11. $|0|$
12. $|^{-}42|$

Simplify.

13. $^{-}|^{-}15|$
14. $|^{-}5| \bullet |^{-}4|$
15. $^{-}|4 \bullet 3|$
16. $^{-}|30 \div 6|$
17. $|^{-}8| + |8|$
18. $|7| - |^{-}7|$

Name ______________________

6.NS.C.6a, 6.NS.C.7

Real-Number Operations with Absolute Value

The **absolute value** of a number is its distance from zero.

$^-|5 - 11| = {^-}|{^-}6| = {^-}6$ $\quad$ $|{^-}4| \bullet |{^-}3| = 4 \bullet 3 = 12$

Simplify.

1. $|{^-}3| =$

2. $|{^-}14| =$

3. $9 + |{^-}4| =$

4. $^-5 \bullet |4| + |5| =$

5. $^-|4 \bullet 7| =$

6. $|21| + 9 =$

7. $23 + |8| =$

8. $|12| - |{^-}15| =$

9. $7 - |{^-}23| + |{^-}7| =$

10. $|{^-}6| \bullet |8| =$

11. $^-|{^-}3 + 7| =$

12. $^-|{^-}5 + 10| \div |{^-}3| =$

13. $|{^-}9| + |23| =$

14. $|{^-}17| \div |{^-}17| =$

15. $|1| - |0| + 6 =$

16. $|{^-}67| - |{^-}17| =$

17. $|4| - |{^-}12 \div 4| =$

18. $|3 - 13| \div |{^-}5| =$

19. $|24| \div |{^-}12| =$

20. $^-|9| \bullet |{^-}9| =$

Name ______________________________

6.NS.C.6a, 6.NS.C.7

Real-Number Operations with Absolute Value

Simplify.

1. $|{}^{-}4 + 8| =$

2. $^{-}|5| \bullet |{}^{-}7| =$

3. $12 \div |{}^{-}3 + 7| =$

4. $^{-}|11| - |{}^{-}20| =$

5. $|6 - 18| + |{}^{-}17 + 9| =$

6. $|{}^{-}27| \div |{}^{-}3| \bullet {}^{-}|{}^{-}2| =$

7. $^{-}48 \div |{}^{-}2 \bullet 6| =$

8. $|{}^{-}83| - |{}^{-}38| =$

9. $|24| \bullet |{}^{-}4 + 1| =$

10. $10 - |{}^{-}13| + |{}^{-}7| =$

11. $|72 \div 8| \bullet |{}^{-}11| =$

12. $^{-}|16 - 21| - |{}^{-}7 + 4| =$

13. $23 + |{}^{-}45 \div 9| - |52| =$

14. $^{-}|{}^{-}4 - 15| =$

15. $|{}^{-}83| + |{}^{-}16| =$

16. $^{-}|4| \bullet |{}^{-}4| =$

17. $|{}^{-}7 + 15| \div |{}^{-}20 - 12| =$

18. $^{-}41 + |{}^{-}29| \bullet |0| =$

19. $|{}^{-}15| \div |3| - |45| =$

20. $|22 \div 11| \bullet |{}^{-}32 \div 8| =$

Name ________________________________ 6.EE.A.2b, 6.EE.A.3

Combining Like Terms

The expression $3x^2 + 2x + 1$ has three terms, $3x^2$, $2x$, and 1. A **term** is either a single number, a variable, or numbers and variables multiplied together. A term in an expression without a variable is called a **constant**, as 1 is above. For terms to be considered "like" terms, they must have the same variable and corresponding variables must have the same exponents. All constant terms are considered "like" terms.

like terms	unlike terms
$3x$ and $8x$	$9y$ and $10z$
$2x^2y$ and $3x^2y$	$3ab$ and $4ab^2$

In the example, $3x$ and $8x$ are like terms with numerical coefficients of 3 and 8. A **numerical coefficient** of a term is simply the number before its corresponding variable. When combining like terms, simply keep the variable the same and combine the numerical coefficients.

$4y + 10y = 14y$ $\quad$ $6b + 9b - 5b = 10b$

$10x - 3x = 7x$ $\quad$ $12x^2y - 10x^2y + 2x^2y = 4x^2y$

Identify the like terms in each problem.

1. $7c + 12c - 2$
2. $19y - 10$
3. $12rt - 10r + 18t$
4. $5r - 10r + 8rs$
5. $5d + 7d - 1$
6. $q + 9 + 2q + 5q$

Simplify. If not possible, write *already simplified.*

7. $8m - 3m$
8. $8y + 12y + 3y$
9. $3s + 8s - 2$
10. $2 + 10k$
11. $8q + 10q + 14$
12. $4 + 8x + 11y$
13. $5a + 6a - 9a$
14. $z + 8m + 4z - 4m$
15. $5w + 2 + w$

Name ________________________________ 6.EE.A.2b, 6.EE.A.3

Combining Like Terms

$$4x + 5y + (^{-}18x) = {}^{-}14x + 5y$$

Combine like terms.

1. $3yz + 5yz =$
2. $3a + 5 + a =$
3. $5x - 5y - 8y + 8x =$
4. $18x + 3x =$
5. $5 - (^{-}4k) =$
6. $7c - 12c =$
7. $13ab + (^{-}12ab) =$
8. $^{-}12x + (^{-}4x) =$
9. $^{-}10n - (^{-}13n) =$
10. $12b + (^{-}34b) =$
11. $4.7x - 5.9x =$
12. $4x^2 + (^{-}8y) + (^{-}3xy) + 5x^2 + 2xy =$
13. $4x + 3y + (^{-}5y) + 3xy + y =$
14. $2x - y + 2x + 3xy =$
15. $5x + 7x =$
16. $23x + 8 + 6x + 3y =$
17. $^{-}e + 8e =$
18. $2xy + 5x + 6xy + 3xy + (^{-}3x) =$
19. $7s + 5x - 8s =$
20. $4xy + 7xy + 6x^2y + 3xy^2 =$

Name ____________________ 6.EE.A.2b, 6.EE.A.3

Combining Like Terms

Combine like terms.

1. $^{-}n + 9n + 3 - 8 - 8n =$

2. $3(^{-}4x + 5y) - 3x(2 + 4y) =$

3. $5 - 4y + x + 9y =$

4. $^{-}2x + 3y - 5x - {}^{-}8y + 9y =$

5. $6(a - b) - 5(2a + 4b) =$

6. $7(x + 5y) + 3(x + 5y) + 5(3x + 8y) =$

7. $12x + 6x + 9x - 3y + (^{-}7y) + y =$

8. $^{-}21x + (^{-}2x) =$

9. $4(x + 9y) - 2(2x + 4y) =$

10. $4(x + 5y) + (5x + y) =$

11. $6x + {}^{-}2y^2 + 4xy^2 + 3x^2 + 5xy^2 =$

12. $^{-}2(c - d) + (c - 3d) - 5(c - d) =$

13. $3x + (^{-}3y) - (4x) + y =$

14. $^{-}3(4x + {}^{-}2y) - 2(x + 3y) - 2(2x + 6y) =$

15. $2b + 3(2b + 8a) - 3(8b + 2a) =$

16. $3[2(^{-}y^2 + y) {}^{-}3] - 3(2x + y) =$

17. $2 \bullet 4x \bullet 3y - 4x \bullet 7y =$

18. $5(3a^2 - 2b^2) + 3a(a + 3b^2) =$

19. $3c + 4d + 2c + 5d - 4c =$

20. $4(x^2 + 3y^2) - y(x^2 + 5y) =$

Name ________________________________ 8.EE.C.7b, HSA-REI.B.3

Solving One-Step Equations (Addition and Subtraction)

When solving equations for a given variable, use the addition or the subtraction property of equality. The addition property means that adding the same number to both sides of an equation will produce an equivalent equation.

$x - 3 = 5$
$x - 3 + 3 = 5 + 3$
$x = 8$

Solve for x by adding 3 to both sides of the equation.

When 3 is added to both sides, $^-3$ and 3 gives a result of 0, leaving x by itself. Therefore, $x = 8$ is the answer.

The subtraction property means that subtracting the same number from both sides of an equation will produce an equivalent equation.

$x + 3 = 5$
$x + 3 - 3 = 5 - 3$
$x = 2$

Solve for x by subtracting 3 from both sides of the equation.

When 3 is subtracted from both sides, $3 - 3$ gives a result of 0, leaving x by itself. Therefore, $x = 2$ is the answer.

Answers can be checked by substituting the value of x back into the equation. This is to make sure the value of x makes a true sentence when substituted into the original equation.

$x - 3 = 5, x = 8$
$8 - 3 = 5$
$5 = 5$

$x + 3 = 5, x = 2$
$2 + 3 = 5$
$5 = 5$

Solve each equation for x. Check your answers.

1. $x - 10 = 23$
2. $6 + x = {}^-11$
3. $^-13 = {}^-6 + x$
4. $7 = 14 + x$
5. $8 = x + 9$
6. $x - 5 = {}^-5$
7. $x + 3 = 12$
8. $7 + x = 7$
9. $7 + x = {}^-7$
10. $^-2 = x - 5$
11. $x + 4 = {}^-15$
12. $9 = x + 12$

Name ____________________

8.EE.C.7b, HSA-REI.B.3

Solving One-Step Equations (Addition and Subtraction)

$$12 + x = {}^{-}24$$
$$12 - 12 + x = {}^{-}24 - 12$$
$$x = {}^{-}36$$

Solve each equation for the given variable.

1. ${}^{-}13 + b = 31$

2. $x - 17 = {}^{-}27$

3. $27 = v + ({}^{-}5)$

4. ${}^{-}4 = x - 3$

5. $12 - ({}^{-}z) = 17$

6. ${}^{-}200 = b + ({}^{-}73)$

7. ${}^{-}13 + x = 18$

8. ${}^{-}w + ({}^{-}7) = {}^{-}56$

9. $3 + x = 9$

10. $z + 3.5 = 4.7$

11. $12 + ({}^{-}g) = 10$

12. $y - 12 = 15$

13. $x + 2 = {}^{-}6$

14. $s - 5 = {}^{-}8$

15. ${}^{-}13 = n + ({}^{-}39)$

16. $r = 4.4 + 3.9$

Name ______________________________ 8.EE.C.7b, HSA-REI.B.3

Solving One-Step Equations (Addition and Subtraction)

Solve each equation for the given variable.

1. $c + 23 = 41$

2. $48 - x = 15$

3. $a + 5.7 = 18.9$

4. $^{-}29 - n = 6$

5. $40 - (^{-}g) = 38$

6. $c - 3 = 4.7$

7. $n + \frac{3}{8} = \frac{5}{8}$

8. $^{-}1.9 + b = ^{-}4.5$

9. $z + (^{-}14) = 13$

10. $^{-}135 + r = ^{-}26$

11. $2\frac{1}{3} + w = 4\frac{2}{9}$

12. $^{-}0.6 - m = ^{-}1.5$

13. $^{-}d + (^{-}61) = 107$

14. $s - \frac{7}{8} = -\frac{3}{4}$

15. $^{-}101 = g - (^{-}28)$

16. $241 + p = ^{-}93$

17. $35.02 - q = 46.1$

18. $5\frac{3}{10} + (^{-}v) = 3.8$

19. $^{-}74 = ^{-}k - (^{-}91)$

20. $^{-}8 + x = 5.62$

Name ____________________ 8.EE.C.7b, HSA-REI.B.3

Solving One-Step Equations (Multiplication and Division)

When solving equations for a given variable, use the multiplication or division property of equality. The multiplication property means that multiplying the same nonzero number by both sides of the equation will produce an equivalent equation.

$\frac{x}{7} = 3$ Multiply both sides of the equation by 7 to isolate x.

$(7)(\frac{x}{7}) = (7)(3)$ Check $\frac{21}{7} = 3$

$x = 21$ $3 = 3$

Therefore, the solution is 21.

The division property means that dividing both sides of the equation by the same nonzero number will produce an equivalent equation.

$^{-}5x = 20$ Divide both sides of the equation by $^{-}5$ to isolate x.

$\frac{^{-}5x}{^{-}5} = \frac{20}{^{-}5}$ Check $^{-}5(^{-}4) = 20$

$x = {^{-}4}$ $20 = 20$

Therefore, the solution is $^{-}4$.

Solve each equation for x. Check your answers.

1. $5x = 35$
2. $18 = {^{-}3x}$
3. $^{-}7x = 49$
4. $-\frac{1}{3}x = 6$
5. $^{-}5x = {^{-}20}$
6. $-\frac{5}{8}x = 10$
7. $\frac{1}{4}x = {^{-}2}$
8. $\frac{2}{3}x = 8$
9. $4 = -\frac{x}{5}$
10. $^{-}4x = 48$
11. $\frac{x}{3} = {^{-}5}$
12. $^{-}6x = 24$

Name ______________________

8.EE.C.7b, HSA-REI.B.3

Solving One-Step Equations (Multiplication and Division)

$3x = 15$

$\frac{3x}{3} = \frac{15}{3}$

$x = 5$

$-\frac{3}{4y} = {}^-6$

$-\frac{4}{3} \bullet -\frac{3}{4y} = {}^-6 \bullet -\frac{4}{3}$

$\frac{1}{y} = 8$ so $y = \frac{1}{8}$

Solve each equation for the given variable.

1. $^-13h = 169$

2. $4b = {}^-36$

3. $10x = {}^-100$

4. $4c = 288$

5. $7x = {}^-63$

6. $4y = {}^-48$

7. $6x = {}^-36$

8. $\frac{8}{k} = \frac{2}{5}$

9. $-({}^-90) = {}^-45z$

10. $^-50 = 2x$

11. $\frac{2}{n} = \frac{1}{9}$

12. $\frac{4}{x} = \frac{2}{9}$

13. $\frac{x}{6} = \frac{6}{9}$

14. $^-35c = 700$

15. $^-4x = {}^-20$

16. $-\frac{x}{6} = \frac{2}{3}$

Name ______________________________

8.EE.C.7b, HSA-REI.B.3

Solving One-Step Equations (Multiplication and Division)

Solve each equation for the given variable.

1. $12.8 = 4b$

2. $-\frac{3}{4} = \frac{n}{16}$

3. $\frac{2}{3}x = {}^{-}10$

4. $-({}^{-}36) = {}^{-}0.25c$

5. $1.6r = 80$

6. $-\frac{x}{8} = \frac{1}{4}$

7. $208x = {}^{-}4$

8. $\frac{2}{k} = \frac{1}{8}$

9. ${}^{-}4.9 = 7m$

10. $0.006w = {}^{-}0.54$

11. $\frac{720}{p} = 9$

12. $-\frac{5h}{50} = {}^{-}0.1$

13. ${}^{-}9s = 6$

14. $15d = -\frac{1}{3}$

15. ${}^{-}({}^{-}27)d = 16.2$

16. $3z = {}^{-}174$

17. $-\frac{4}{5j} = {}^{-}8$

18. $\frac{56}{63} = -\frac{x}{9}$

Name ______________________________ 8.EE.C.7b, HSA-REI.B.3

Solving Basic Multistep Equations

When solving equations for a given variable, sometimes you need to use more than one of the properties of equality.

$3x - 2 = 7$

$3x - 2 + 2 = 7 + 2$ Add 2 to both sides of the equation.

$\frac{3x}{3} = \frac{9}{3}$ Now, divide both sides of the equation by 3.

$x = 3$ Check $3(3) - 2 = 7$

$9 - 2 = 7$

$7 = 7$

Therefore, the solution is 3.

Here are some steps to follow when solving multi-step equations.

1. Simplify both sides of the equation (if possible).
2. Use the addition or subtraction property of equality to isolate terms containing the variable.
3. Use the multiplication or the division property of equality to further isolate the variable.
4. Check the solution.

Solve each equation for x. Check your answers.

1. $6x - 3 = 21$
2. $^{-}6 + \frac{x}{4} = 1$
3. $18 - 3x = {}^{-}12$
4. $7 + 2x = {}^{-}13$
5. $^{-}4 = 7x + 8 - 8x$
6. $13 = 9 - \frac{x}{5}$
7. $^{-}7 - x = {}^{-}5$
8. $5x + 9 - 4x = 12$
9. $^{-}8x - 13 = 19$
10. $^{-}3 = {}^{-}5 - 2x$
11. $\frac{1}{2}x + 9 = 15$
12. $^{-}7 = 3x - 15 - 7x$

Name ____________________

8.EE.C.7b, HSA-REI.B.3

Solving Basic Multistep Equations

$$4x + 4 = 12$$
$$4x + 4 - 4 = 12 - 4$$
$$4x = 8$$
$$x = 2$$

Solve each equation for the given variable.

1. $7x - 12 = 2$
2. $7a - 4 = 24$
3. $4b - 7 = 37$
4. $3c - 9 = 9$
5. $8 - 9y = 35$
6. $8 - 12x = 32$
7. $1.3x + 5 = {}^{-}5.4$
8. $3(y + 4) + 5 = 35$
9. $0 = 25x + 75$
10. $3 - \frac{1}{5}e = {}^{-}7$
11. $5 - \frac{1}{2}x = {}^{-}9$
12. $2x = 6 + ({}^{-}18)$
13. $7 - \frac{1}{9}k = 32$
14. $\frac{3}{12}w + 2 = 11$
15. $\frac{2x}{5} + 3 = 9$
16. $\frac{x}{3} - 8 = {}^{-}12$
17. $5(e + 5) = {}^{-}10$
18. $8 - \frac{1}{2}y = {}^{-}6$

Name______________________________ 8.EE.C.7b, HSA-REI.B.3

Solving Basic Multistep Equations

Solve each equation for the given variable.

1. $5z - 8 = {}^{-}28$
2. $4k + 7 = {}^{-}9$
3. $13x + 7 = {}^{-}32$
4. $2x + 12 = 6$
5. $7.2 + 4x = 19.2$
6. $2(w - 6) = 8$
7. $7h + 1 = {}^{-}13$
8. $3(c - 2) = 15$
9. $6x - 5 = {}^{-}41$
10. ${}^{-}3 + 2n = {}^{-}15$
11. $5e + ({}^{-}9) = 26$
12. $\frac{m}{3} - 7 = {}^{-}10$
13. $6x - 2 = 34$
14. ${}^{-}8(r - 2) = 40$
15. $5n - 8 = {}^{-}23$
16. $2 + (\frac{1}{5})x = {}^{-}7$
17. $5 - (\frac{1}{2})g = 12$
18. $3x - 4 = 14$
19. ${}^{-}6 = \frac{3z}{4} + 12$
20. $2(f + 7) - 8 = 22$
21. $4.7 = {}^{-}3.4m - 5.5$
22. $32 = \frac{4}{6}x - 34$

Name ______________________________

8.EE.C.7b, HSA-REI.B.3

Solving Equations with Variables on Both Sides

Often, equations with a variable on both sides of the equation need to be solved. This requires one additional step of getting the variable on one side only.

$$4x - 1 = 2x + 7$$

$$4x - 2x - 1 = 2x - 2x + 7$$ Get the variable on one side of the equation.

$$2x - 1 = 7$$ Now, solve for x using the properties of equality.

$$2x - 1 + 1 = 7 + 1$$

$$\frac{2x}{2} = \frac{8}{2}$$

$$x = 4$$

Check $4(4) - 1 = 2(4) + 7$

$$16 - 1 = 8 + 7$$

$$15 = 15$$

Therefore, the solution is 4.

Combine the variables on the side of the equation with the greater variable coefficient, in order to avoid solving an equation with a variable which has a negative coefficient.

Solve each equation for x. Check your answers.

1. $9x - 12 = 3x$
2. $8x - 12 = 15x - 4x$
3. $11 + 6x = 2x - 13$
4. $^{-}5x = 9 - 2x$
5. $^{-}8x - 10 = 4x + 14$
6. $10x - 5 = 21 - 3x$
7. $^{-}12x = 14 - 5x$
8. $19 - 3x = 21 + x$
9. $4x + 12 = {}^{-}3x - 6 + 4x$
10. $14x + 5 = 8x - 1$
11. $7x - 3 = -4x - 25$
12. $^{-}9x + 5 = {}^{-}22 - 6x$

Name ____________________

8.EE.C.7b, HSA-REI.B.3

Solving Equations with Variables on Both Sides

$$
\begin{aligned}
6x - 7 &= x + 33 \\
6x - x - 7 &= x - x + 33 \\
5x - 7 + 7 &= 33 + 7 \\
5x &= 40 \\
x &= 8
\end{aligned}
$$

Solve each equation for the given variable.

1. $4x - 6 = 8 - 3x$

2. $2(x + 3) = 12 - x$

3. $5(5 - z) = 4(z - 5)$

4. $4e + 6 = {}^{-}8 + 11e$

5. ${}^{-}4(x - 6) = 2(7 - 7x)$

6. $2m - 9 = 8m - 27$

7. $b + 9 = 6 + 2b$

8. ${}^{-}9j + 3 = {}^{-}32 - 4j$

9. $5(j - 4) = {}^{-}8 - j$

10. $8({}^{-}9x + 4) = {}^{-}3(6x + 9) + 5$

11. $3d - 5 = {}^{-}9 + 2d$

12. ${}^{-}2 + 6d = 34 - 3d$

13. $3(k + 4) = {}^{-}3 - 2k$

14. $5(m - 3) = 27 - 2m$

15. ${}^{-}j + 2 = {}^{-}14 - 5j$

16. ${}^{-}(x + 7) - 5 = 4(x + 3) - 6x$

17. $5(x - 1) = 2x + 4(x - 1)$

18. ${}^{-}3(5k + 5) = 54 - 12k$

19. $3e + 1 = 36 - 4e$

20. ${}^{-}6r + 12 = {}^{-}2 + 8r$

Name ____________________ 8.EE.C.7b, HSA-REI.B.3

Solving Equations with Variables on Both Sides

Solve each equation for the given variable.

1. $7 - 6a = 6 - 7a$

2. $3c - 12 = 14 + 5c$

3. $3x - 3 = {}^{-}3x + {}^{-}3$

4. $2x - 7 = 3x + 4$

5. $9a + 5 = 3a - 1$

6. $8(x - 3) + 8 = 5x - 22$

7. $5d + 7 = 4d - 9$

8. ${}^{-}10w + 6 = {}^{-}7w + {}^{-}9$

9. ${}^{-}7c + 9 = c + 1$

10. $2n + 6 = 5n - 9$

11. $\frac{5}{2}x + 3 = \frac{1}{2}x + 15$

12. $5 + 3x = 7(x + 3)$

13. $12m - 9 = 4m + 15$

14. $2(f - 4) + 8 = 3f - 8$

15. ${}^{-}6 - ({}^{-}2n) = 3n - 6 + 5$

16. $4(2y - 4) = 5y + 2$

17. $2(r - 4) = 5[r + ({}^{-}7)]$

18. $6(x - 9) = 4(x - 5)$

19. $4(z + 5) - 3 = 6z - 13$

20. $4e - 19 = {}^{-}3(e + 4)$

21. $4(-6x - 2) = -22x$

22. $-\frac{1}{3}x - 5 = 7 - x$

Name ____________________ 6.EE.A.2, 7.EE.B.3, 8.EE.C.7b

Writing and Solving Equations

The sum of three times a number and 25 is 40. Find the number.

$$3x + 25 = 40$$
$$3x + 25 - 25 = 40 - 25$$
$$3x = 15$$
$$x = 5$$

The number is 5.

Write an equation for each word problem and solve it.

1. The difference of a number and $^{-}3$ is 8. Find the number.

   Equation ____________________ **Solution** ____________________

2. Twice a number added to 9 is 15. Find the number.

   Equation ____________________ **Solution** ____________________

3. Twelve subtracted from 3 times a number is 15. Find the number.

   Equation ____________________ **Solution** ____________________

4. The sum of 4 times a number and 5 is $^{-}7$. Find the number.

   Equation ____________________ **Solution** ____________________

5. The product of a number and 5 is 60. Find the number.

   Equation ____________________ **Solution** ____________________

6. The difference of 5 times a number and 6 is 14. Find the number.

   Equation ____________________ **Solution** ____________________

7. The sum of a number and $^{-}6$ is 10. Find the number.

   Equation ____________________ **Solution** ____________________

8. The quotient of a number and 4 is $^{-}12$. Find the number.

   Equation ____________________ **Solution** ____________________

Name ______________________

6.EE.A.2, 7.EE.B.3, 8.EE.C.7b

Writing and Solving Equations

Write an equation for each word problem and solve it.

1. Six times the difference of a number and 9 is 54. Find the number.

   Equation ______________________ **Solution** ______________________

2. The sum of 8 times a number and 3 is 59. Find the number.

   Equation ______________________ **Solution** ______________________

3. The sum of 5 times a number and $^{-}11$ is $^{-}16$. Find the number.

   Equation ______________________ **Solution** ______________________

4. Twelve times the sum of a number and $^{-}8$ is 48. Find the number.

   Equation ______________________ **Solution** ______________________

5. The sum of 5 times a number and 2 is $^{-}13$. Find the number.

   Equation ______________________ **Solution** ______________________

6. The sum of 7 times a number and 11 is 81. Find the number.

   Equation ______________________ **Solution** ______________________

7. Three times the sum of a number and $^{-}2$ is $^{-}15$. Find the number.

   Equation ______________________ **Solution** ______________________

8. Five times the sum of a number and 2 is 35. Find the number.

   Equation ______________________ **Solution** ______________________

Name ____________________

6.EE.A.2, 7.EE.B.3, 8.EE.C.7b

Writing and Solving Equations

Write an equation for each word problem and solve it.

1. A number increased by 10 is 38. Find the number.

   Equation ____________________ **Solution** ____________________

2. A number times 8 is $^{-}96$. Find the number.

   Equation ____________________ **Solution** ____________________

3. Thirty-two is 7 less than 3 times a number. Find the number.

   Equation ____________________ **Solution** ____________________

4. A number multiplied by 4 and increased by 7 is $^{-}25$. Find the number.

   Equation ____________________ **Solution** ____________________

5. A number decreased by 12 is the same as 3 times the number. Find the number.

   Equation ____________________ **Solution** ____________________

6. Four times the sum of a number and 7 is 44 less than the number. Find the number.

   Equation ____________________ **Solution** ____________________

7. A number increased by 16 is the same as 8 times the sum of the number and 9. Find the number.

   Equation ____________________ **Solution** ____________________

8. Four times a number is decreased by 9 and then increased by 12. The result is 5 less than 2 times the number. Find the number.

   Equation ____________________ **Solution** ____________________

Name ____________________ 6.EE.B.8, HSA-REI.B.3

Solving Inequalities

$$^{-}10n + 5 \le 55$$
$$^{-}10n + 5 - 5 \le 55 - 5$$
$$^{-}10n \le 50$$
$$n \ge {}^{-}5$$

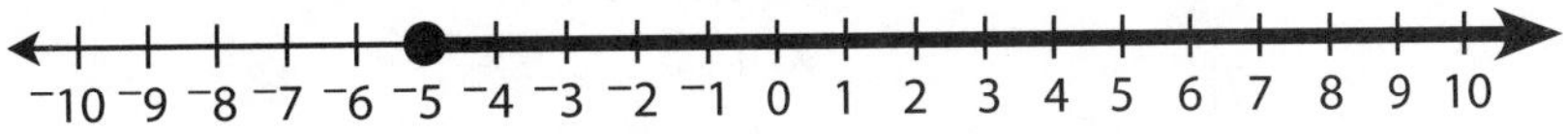

Solve each inequality and graph its solution set.

1. $^{-}4(3d + 2) \le 4$

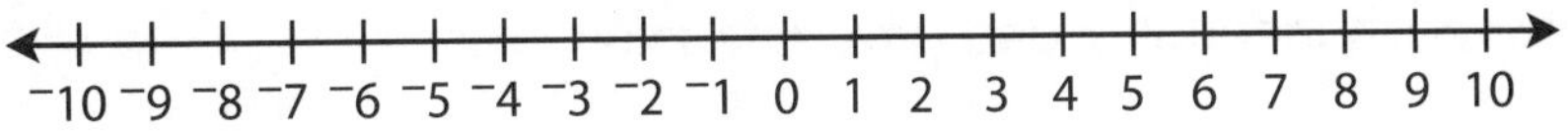

2. $10 - 5x - 20 \ge {}^{-}20$

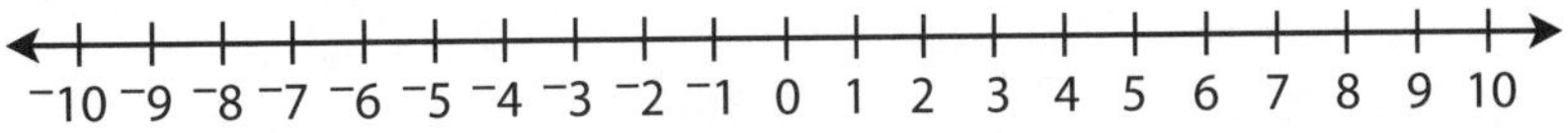

3. $^{-}15 > 4y - 7 - 3y - 4$

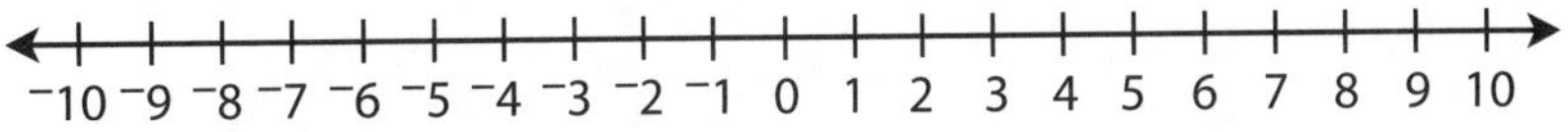

4. $4x - 7 < 9$

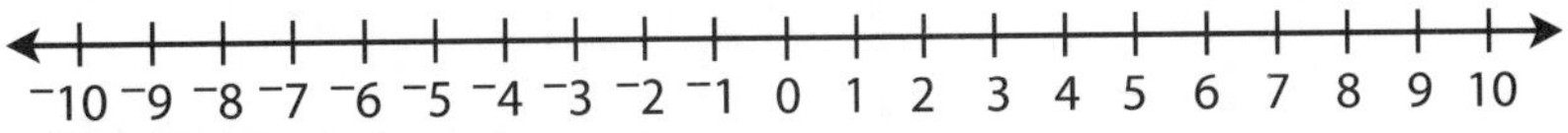

5. $6n - 3 > 33$

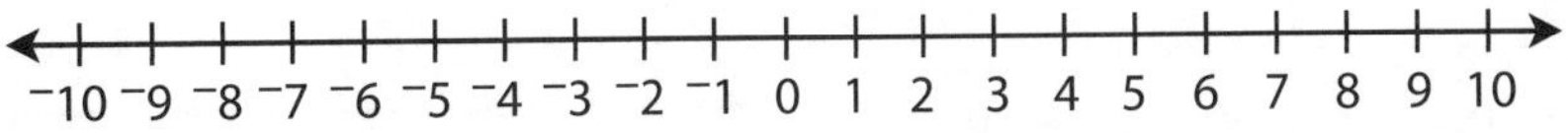

6. $3(3c - 4) \le 15$

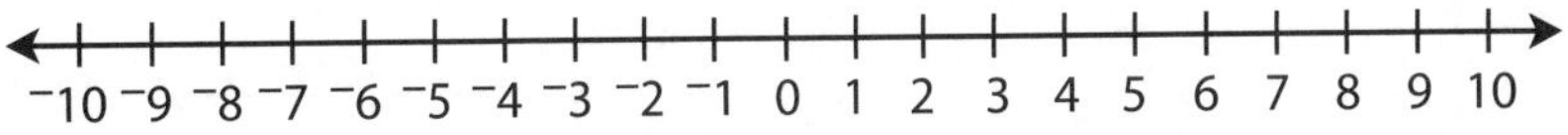

7. $5z - 1 > 9$

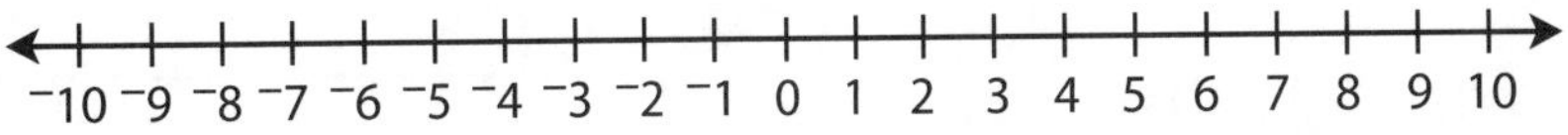

8. $5 > 4x - 11$

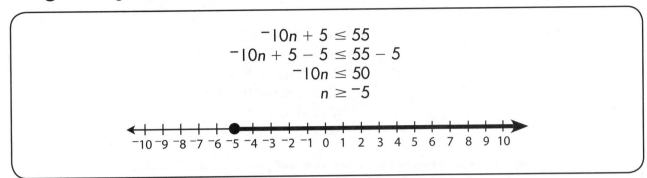

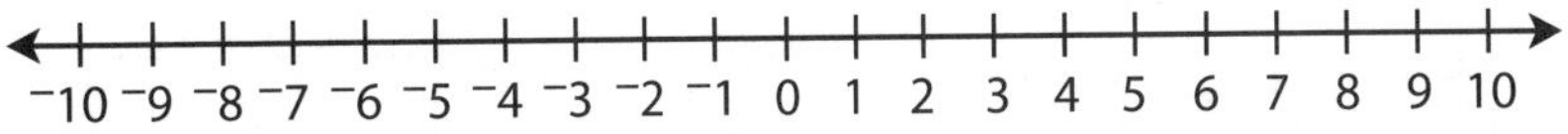

Name ____________________________

6.EE.B.8, HSA-REI.B.3

Solving Inequalities

$$-10x > 4x - 42$$
$$-10x + 10x > 4x + 10x - 42$$
$$0 > 14x - 42$$
$$0 + 42 > 14x - 42 + 42$$
$$42 > 14x$$
$$x < 3$$

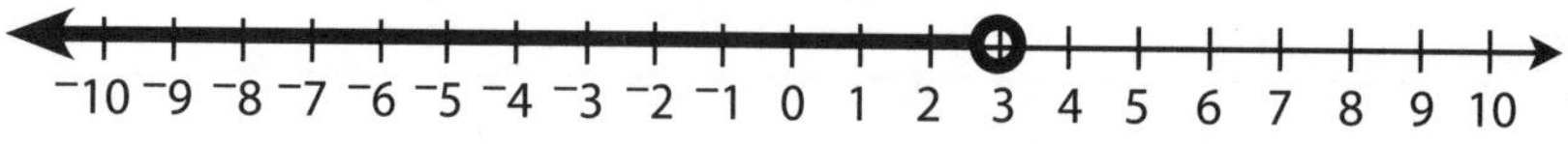

Solve each inequality and graph its solution set.

1. $-2a < 10 + 3a$

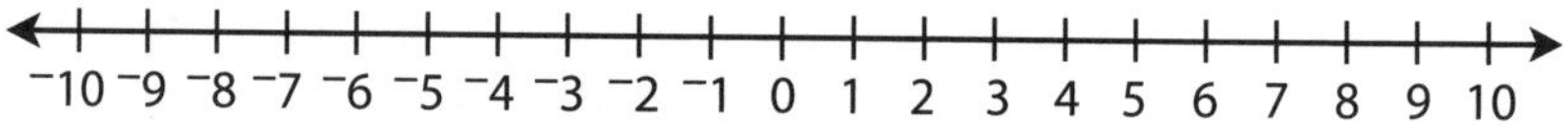

2. $4w > 2w + 8$

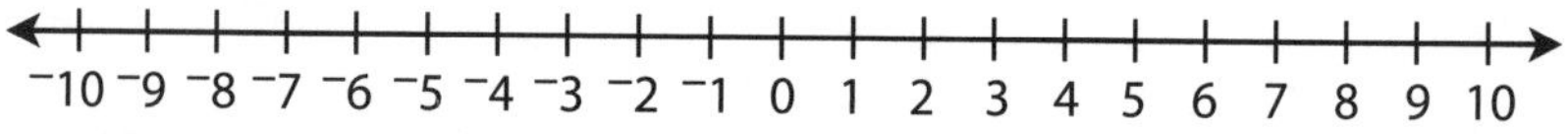

3. $2(k + 4) \leq 3(2k - 4)$

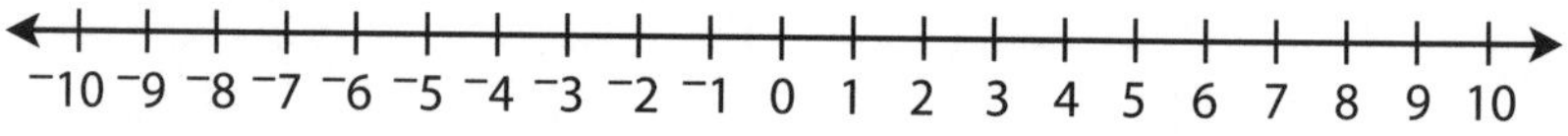

4. $5n + 3 \geq -12$

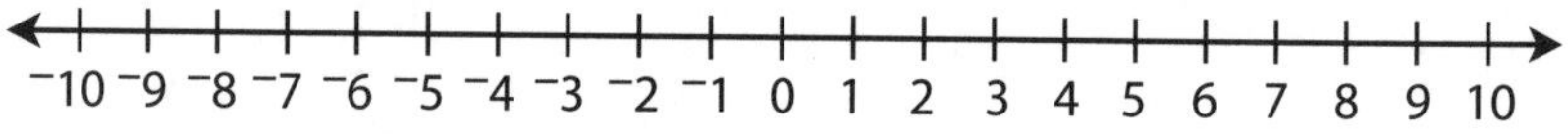

5. $5c + 2 < 2c + (-7)$

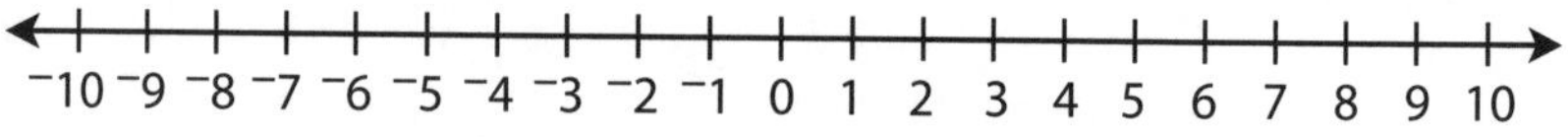

6. $5x - 20 > 2x + 1$

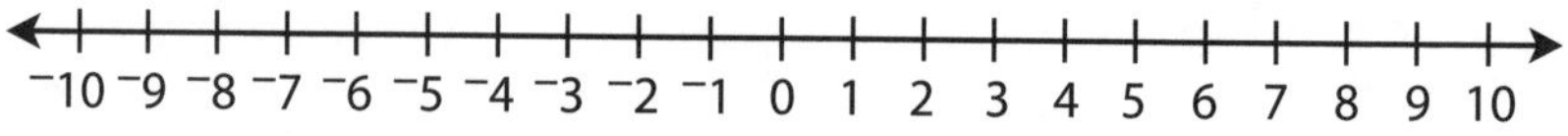

7. $3(s - 4) \geq 4s - 12$

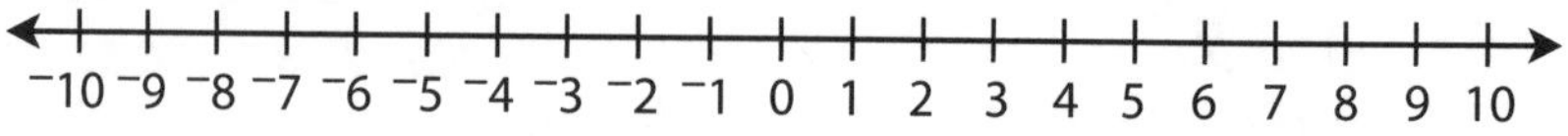

8. $-9 - e > 3e + 11$

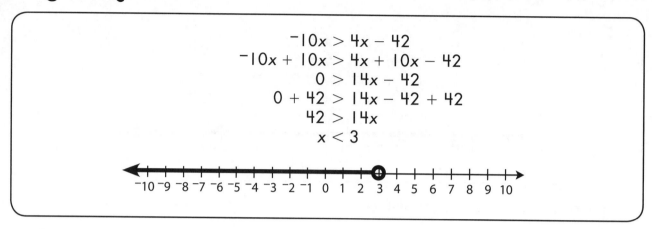

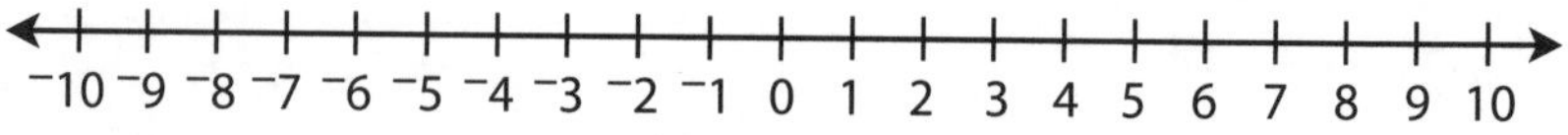

Name ______________________________ 6.EE.B.8, HSA-REI.B.3

Solving Inequalities

Solve each inequality and graph its solution set.

1. $15e - 3 \leq 20e + 17$

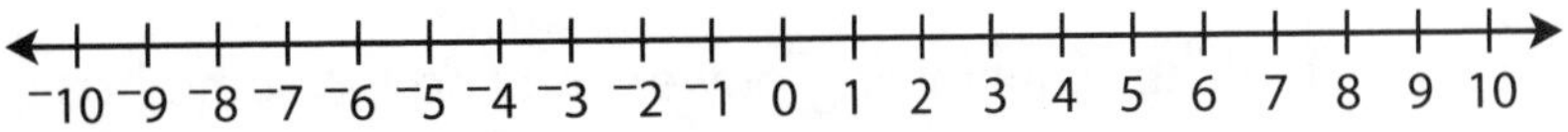

2. $6x - 4 > 2(x - 6)$

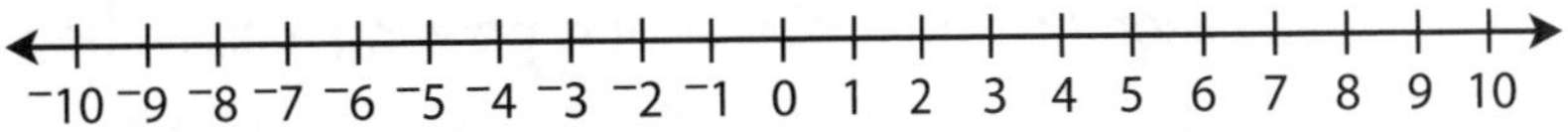

3. $7c - 8 \geq 6$

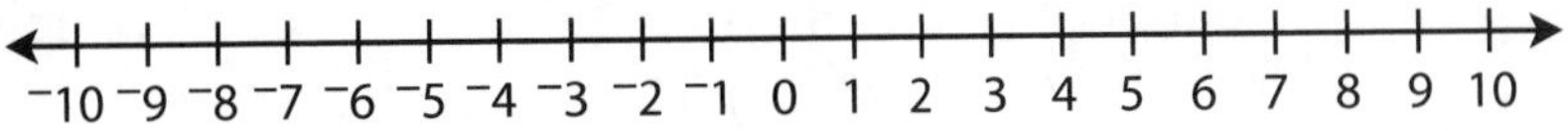

4. $5x + (^{-}3) \geq 2(3 + x)$

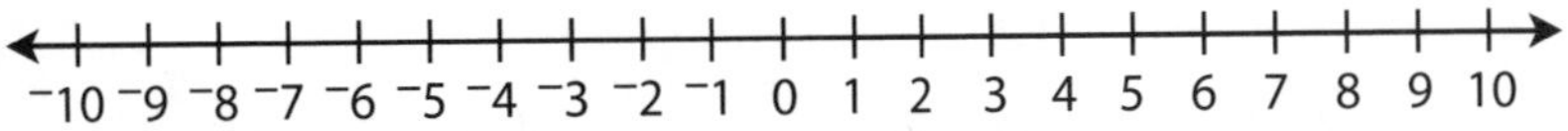

5. $^{-}8 < 2(2 + 3r)$

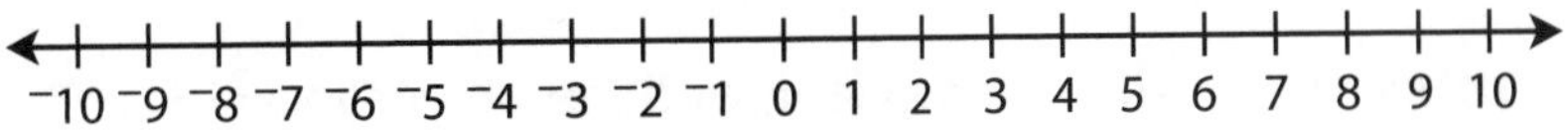

6. $6d < 3d - 18$

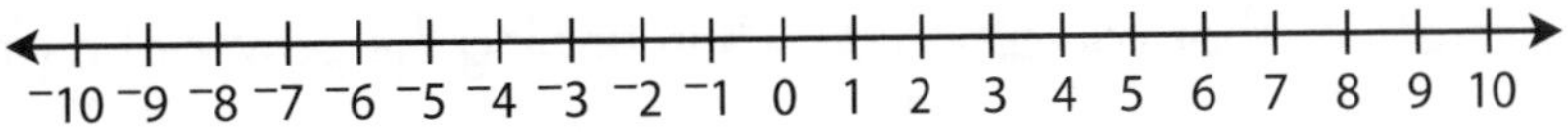

7. $7m + 9 \leq 6(m + 3)$

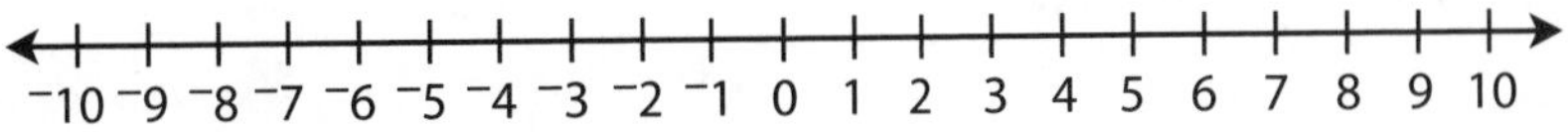

8. $3(2x + 4) \geq 7x + 8$

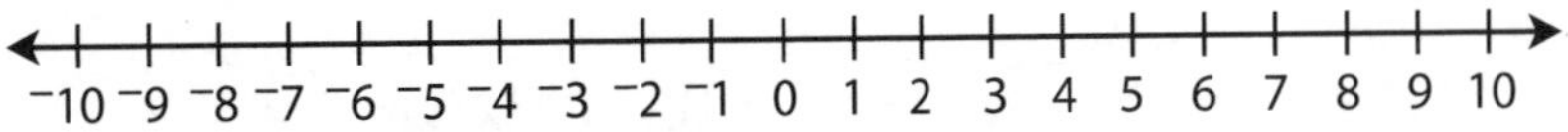

9. $^{-}4.2 > 0.6y$

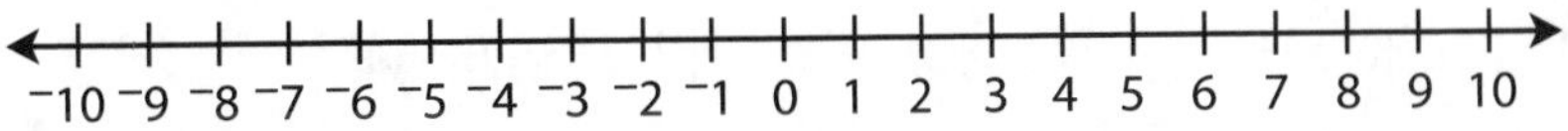

10. $\frac{a}{4} + 3 \leq 5$

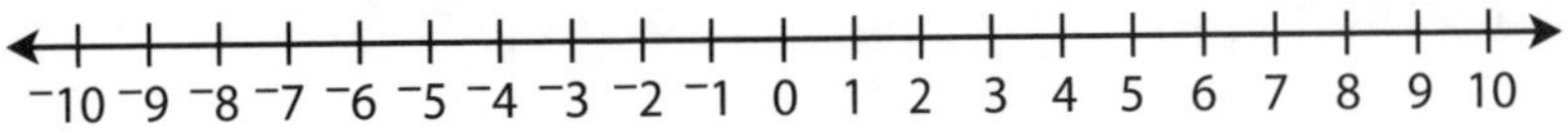

Name ____________________ 8.EE.A.1

Multiplication and Powers of Exponents

When multiplying exponents, it is important to remember the following properties:

1. When multiplying powers having the same base, add the exponents, keeping the same base. (Remember: in a^3, a is the base, 3 is the exponent, and a^3 is the power.) For example, $x^3 \bullet x^5 = x^{3+5} = x^8$.
2. When finding a power of a power, multiply the exponents. For example, $(x^3)^2 = x^6$.
3. When finding the power of a product, find the power of each factor and multiply. For example, $(x \bullet y)^2 = x^2 \bullet y^2$.

Simplify $(5x^3)^2(xy)^3$
$(5^2x^6)(x^3y^3)$
$= 25x^9y^3$

Simplify $(2x^4)^3(^-x^2)^3$
$(2^3x^{12})(^-x^6)$
$= {}^-8x^{18}$

1. What do you do with the exponents when multiplying powers that have the same base?

2. Label the base, the exponent, and the power in x^3.

3. Explain what you are to do when finding the power of a product.

Simplify each expression.

4. $3x \bullet x^2$

5. $(^-6xy)^2(x^2y)^3$

6. $(8xy)^2$

7. $^-4x^4 \bullet x^3$

8. $(^-3x^2y)^3$

9. $(4c^2)(^-5c^7)$

10. $(^-x^2)(^-x)^2$

11. $^-xy(^-xy)^2$

12. $(3x^3)(5x^5)$

13. $(x^3y^3)^3$

Name ______________________________ 8.EE.A.1

Multiplication and Powers of Exponents

Rule: $(x^a)^b = x^{ab}$	Example: $(x^2y^3)^3 = x^6y^9$
Rule: $x^a \bullet x^b = x^{a+b}$	Example: $x^3 \bullet x^5 = x^8$

Simplify each expression.

1. $c \bullet c^2 \bullet c^3 =$

2. $e \bullet e^2 \bullet e^3 \bullet e^4 \bullet e^5 =$

3. $a^3 \bullet a^4 \bullet a^7 \bullet a =$

4. $(3xy^2)(2x^2y^3) =$

5. $(2a^2b)(4ab^2) =$

6. $(5f)(^-3f^3)(2f)^2 =$

7. $(m^2n)^3(4mn^2)(mn) =$

8. $(4k^2)(^-3k)(3k^5) =$

9. $(^-2c^4)(4cd)(^-cd^2) =$

10. $(3x^3)^2(3x^4)(^-3x^2)^3 =$

11. $(^-1)(x)(^-x^2)(x^3)(^-x^2) =$

12. $(3x^2)(^-3x^5) =$

13. $(c^2h)^2(3ch^3)(2c^3h^4)^2 =$

14. $(^-4p^3)(^-4p^6)(^-2p^9) =$

15. $(12c^3)(2g^3)(4ch)^3 =$

16. $(4x^2y^3)^3(x^3y)(^-x^2y^2) =$

17. $(^-4f^3)^4(^-3m^3)^5 =$

18. $(2c^2d^2)^4(^-5cd^4) =$

19. $(3x)^2(2x^3y^6)(^-5x^6y^2) =$

20. $(3x)(^-4y^2)(6x^3y) =$

Name ____________________ 8.EE.A.1

Multiplication and Powers of Exponents

Simplify each expression.

1. $(^-4xy^3)^3 =$
2. $(x^2y^3)(x^3y) =$
3. $(^-6x^4y^6)^3 =$
4. $(5x^2y^4)^3 =$
5. $(6x^5y^4)(5x^5y^4) =$
6. $(2x)^4(x^3y)^2 =$
7. $(4x^3y^2)^3(^-2x^2y^4) =$
8. $(3x^2y)(^-8xy^4) =$
9. $(^-2x^2y)^4 =$
10. $(x^2y^3)^2(x^3y^2)^4 =$
11. $(^-4x^3y^3)^4(^-8x^3)^2 =$
12. $(3xy^3)(^-4x^2y^4)^2(xy^3) =$
13. $(^-3x^3y)^3 =$
14. $(^-2x^4y^5)^3 =$
15. $(3x^2y^3)^4(x^3y)(7xy^3) =$
16. $(6x^2y^3)(4xy^2)^3(3x^2y)^2 =$
17. $(^-2x^3y^3z)^4(2xyz^4)^2 =$
18. $(5xy^3)(^-5xy^2)^5 =$
19. $(^-3x^2y^3)^2 =$
20. $(4z^4)^2(2x^2y)(^-3xy^3z^5) =$

Name ________________________________ 8.EE.A.1

Division of Exponents

When dividing exponents, it is important to remember the following properties:

1. When dividing powers that have the same base, subtract the exponents.

   For example, $\frac{x^4}{x^2} = x^{4-2} = x^2$, where x cannot be equal to 0.

2. When finding a power of a quotient, find the power of the numerator and the power of the denominator and divide.

   For example, $(\frac{x}{y})^3 = \frac{x^3}{y^3}$, where y cannot be equal to 0.

Simplify: $\frac{6^8}{6^6}$ $\quad (\frac{3}{4})^{-2}$

$= 6^{8-6}$ $\quad = \frac{3^{-2}}{4^{-2}}$

$6^2 = 36$ $\quad = \frac{4^2}{3^2} = \frac{16}{9}$

1. Explain what you do with the exponents when dividing powers that have the same base.

Evaluate each expression.

2. $\frac{5^6}{5^3}$ 3. $\frac{(-3)^2}{3^2}$ 4. $\frac{3^5}{3^2}$ 5. $\frac{5^4 \bullet 5}{5^7}$ 6. $\frac{8^6}{8^3}$

7. $\frac{7^3}{7}$ 8. $\frac{4^8}{4^8}$ 9. $\frac{6^4 \bullet 6^3}{6^5}$ 10. $\frac{4^6}{4^2}$ 11. $\frac{10^5}{10^3}$

Simplify each expression.

12. $\frac{x^5}{x^3}$ 13. $x^7 \bullet \frac{1}{x^5}$ 14. $\frac{18x^4y^4}{6x^2y^4}$

15. $\frac{4x^4y^4}{4x^2y}$ 16. $\frac{6x^2y^4}{3y^2}$ 17. $x^4 \bullet \frac{1}{x^2}$

Name ________________________________ 8.EE.A.1

Division of Exponents

Rule: $\frac{x^a}{x^b} = x^{a-b}$ Example: $\frac{x^6}{x^4} = x^{6-4} = x^2$ $\frac{x^3}{x^{-2}} = x^{3-(^-2)} = x^5$

Simplify each expression.

1. $\frac{^-12m^5}{6m} =$

2. $\frac{x^3}{x^2} =$

3. $\frac{9a^3b^5}{^-3ab^2} =$

4. $\frac{16c^3}{^-4c^2} =$

5. $\frac{d^3}{d^2} =$

6. $\frac{^-3p^8}{6p^2} =$

7. $\frac{^-54c^2d^4}{^-8cd} =$

8. $\frac{49r^{13}}{^-7r^8} =$

9. $\frac{45k^7r^3}{^-3k^5} =$

10. $\frac{^-14c^{15}d^3}{^-2c^9d} =$

11. $\frac{(5k)(^-8k^5)}{10k^3} =$

12. $\frac{24x^2y}{^-4x^2} =$

13. $\frac{4x^2y^3z^4}{2xy^2z^3} =$

14. $\frac{9a^{11}}{a^3} =$

15. $\frac{22y^5z^8}{2yz^7} =$

16. $\frac{b^{14}c^9}{b^5c^4} =$

Name ____________________ 8.EE.A.1

Division of Exponents

Simplify.

1. $\frac{a^5}{a^3} =$

2. $\frac{a^5b^2}{2a^2} =$

3. $\frac{13m^6n^7}{39m^3n^5} =$

4. $\frac{9x^8y^7z^8}{18x^5y^5z^4} =$

5. $\frac{^-2c^2d^3}{^-8cd^2} =$

6. $\frac{10a^6b^8}{40a^2b^2} =$

7. $\frac{18a^6b^2c^6}{36a^4bc^2} =$

8. $\frac{5x^3y^2z^2}{5x^2yz} =$

9. $\frac{45x^9y^{10}z^5}{51x^9y^8z^3} =$

10. $\frac{16x^2y^4}{4x^2y^3} =$

11. $\frac{18x^6y^3z^4}{12x^3y^2z^3} =$

12. $\frac{72x^5y^5z^6}{8x^4yz^3} =$

13. $\frac{18a^9b^3}{12a^2b^2} =$

14. $\frac{44x^8y^2}{11x^7y} =$

15. $\frac{(6x^3)(3x^8)}{^-12x^{10}} =$

16. $\frac{21k^9}{(3k)(7k^4)} =$

17. $\frac{(110c^3)\,(^-c^9)}{11c^5} =$

18. $\frac{(3xy)(4x^2y)}{^-6xy^2} =$

Name ______________________ 8.EE.A.1

Negative and Zero Exponents

Given a nonzero number *a* and a positive integer *n*, the following definitions of negative exponents and zero exponents are stated.

1. For a negative exponent: the expression a^{-n} is the reciprocal of a^n. This is written: $a^{-n} = \frac{1}{a^n}$, where $a \neq 0$.
2. For a zero exponent: any nonzero number raised to the 0 power will have an answer of 1. This is written: $a^0 = 1$, where $a \neq 0$.

Evaluate: 5^{-3}

$= \frac{1}{5^3}$

$= \frac{1}{125}$

Simplify by rewriting with positive exponents. $2x^{-3}y^2z^{-4}$

$= 2 \cdot \frac{1}{x^3} \cdot y^2 \cdot \frac{1}{z^4}$

$= \frac{2y^2}{x^3z^4}$

Note: When rewriting expressions to positive exponent form, simply move the factors from the denominator to the numerator or vice versa, leaving out the in-between step.

1. Any number raised to the zero power has what value?

Evaluate each expression.

2. 4^{-2}

3. $7^3 \cdot 7^{-3}$

4. $3 \cdot 3^{-1}$

5. $-5^0 \cdot \frac{1}{3^{-3}}$

6. $(6^2)^{-2}$

7. $(-2^{-3})^{-1}$

Rewrite each expression with positive exponents.

8. x^{-8}

9. $\frac{1}{3x^{-2}}$

10. $(-3)^0x^{-2}$

11. x^{-10}

12. $x^{-3}y^4$

13. $\frac{6}{x^{-2}}$

Name ____________________ 8.EE.A.1

Negative and Zero Exponents

Rule: $x^{-a} = \frac{1}{x^a}$ Example: $4^{-2} = \frac{1}{16}$ Example: $4x^{-2} = \frac{4}{x^2}$ Example: $(2x)^{-3} = \frac{1}{8x^3}$

$4^{-2} = \frac{1}{4^2} = \frac{1}{16}$ $\frac{1}{(2x)^3} = \frac{1}{8x^3}$

Rule: $x^0 = 1$

Simplify.

1. $4cd^{-5}$
2. 3^0
3. $3a^4b^{-3}$
4. 4^{-5}
5. $(^-2)^{-2}$
6. $(3xy)^{-1}$
7. $(3x)^{-3}$
8. $7x^{-3}$
9. $^-2x^{-3}$
10. $(6y^2)^{-2}$
11. $(\frac{4}{5})^{-2}$
12. $4m^3n^{-5}$
13. $(^-11x^3y)^{-2}$
14. $(c^0d)^{-2}$
15. $(\frac{x^2}{y^3})^{-2}$
16. $(\frac{2}{3})^{-1}$
17. b^0
18. c^{-7}

Name ______________________ 8.EE.A.1

Negative and Zero Exponents

Simplify.

1. $6x^{-3}$

2. $4x^{-5}$

3. $\frac{1}{5x^{-4}}$

4. $4x^{-4}y^{-2}$

5. $\frac{1}{(2x)^{-3}}$

6. $(3x^{-2})^2$

7. $14x^{-8}y$

8. $(^-5x^3)^{-2}$

9. $(^-3x^{-2}y^5z^3)^{-2}$

10. $8x^{-2}y^{-4}$

11. $2x^0y^{-5}z^2$

12. $(4x^3y^{-4})^2$

13. $(\frac{x^3}{2y^{-2}})^{-3}$

14. $4m^{-5}n^{-1}$

15. $(\frac{x^2}{y^3})^{-2}$

16. $(^-14c^2d)^{-1}$

17. $9xy^{-4}$

18. $(^-6s^3p^2)^{-3}$

Name ____________________

7.EE.A.1, HSA-APR.A.1

Adding and Subtracting Polynomials

When adding polynomials, the following two methods can be used.

1. The horizontal method:

   Add $(3x^4 + 8x^2 - 2)$ and $(^-4x^2 + 7x^4 + 5)$.
   First group like terms in descending order. $(3x^4 + 7x^4) + (8x^2 - 4x^2) + (^-2 + 5)$
   Simplify. $10x^4 + 4x^2 + 3$

2. The vertical method:

   Add $(4x^7 + 6x^4 - 10 + 2x^5)$ and $(10x^5 - 5x^7 - 3x^4 + 12)$.
   Group like terms in columns in descending order.

   $$\begin{array}{r} 4x^7 + 2x^5 + 6x^4 - 10 \\ + \quad ^-5x^7 + 10x^5 - 3x^4 + 12 \\ \hline ^-x^7 + 12x^5 + 3x^4 + 2 \end{array}$$

   Simplify.

When subtracting polynomials, change the sign of each term in the second polynomial and then choose from one of the methods above to simplify.

Subtract. $(10x^7 - 4x^4 + 7x^2 + 2) - (12x^4 + 6x^7 + 8 - 5x^2)$

Change sign of each term in second polynomial. $10x^7 - 4x^4 + 7x^2 + 2 - 12x^4 - 6x^7 - 8 + 5x^2$

Choosing method 1 above, group like terms in descending order.

$(10x^7 - 6x^7) + (^-4x^4 - 12x^4) + (7x^2 + 5x^2) + (2 - 8)$

Simplify. $4x^7 - 16x^4 + 12x^2 - 6$

1. What is important to remember to do when subtracting polynomials?

Simplify.

2. $(7x - 5) - (3x + 7)$

3. $(2x^2 - 3x + 1) - (5x^2 - 3x + 10)$

4. $(6x + 7) + (8x - 3)$

5. $(9x^2 + 3x - 5) - (3x^2 + 4x - 10)$

6. $(8x^4 - 9x^3 + 2x) - (6x^4 + 7x^3 - 3x)$

7. $(x^3 - x^2 + 2) - (x^3 + x^2 + 5)$

Name______________________________ 7.EE.A.1, HSA-APR.A.1

Adding and Subtracting Polynomials

$$(x^2 + 4x + 2) - (2x^2 + 7x - 6) = {}^{-}x^2 - 3x + 8$$

Add or subtract the polynomials by combining like terms.

1. $(4y^2 - 9y) - ({}^{-}5y^2 + 8y - 8) =$

2. $(6x^2 + 2x + 6) - (4x^2 - 2x + 3) + ({}^{-}5x^2 + 5x + 6) =$

3. $({}^{-}2x^3 + 3x^2 + 9) + ({}^{-}8x^3 - 2x^2 + {}^{-}4x) =$

4. $(2x^2 - 9x - 8) - (2x^3 - 7x^2 + {}^{-}2) =$

5. $(4x^3 - 2x^2 - 12) + (6x^2 + 3x + 8) =$

6. $(3x^4 - 3x + 1) - (4x^3 - 4x - 8) =$

7. $(7x^2 - x - 5) - (3x^2 - 3x + 5) =$

8. $(x^3 - x^2 + 3) - (3x^3 - x^2 + 7) =$

9. $({}^{-}2x^2 + 4x - 12) + (5x^2 - 5x) =$

10. $(4x^3 - 5x^2 - 9) - (6x^3 - 5x - 4) - (5x^3 - 4x^2 - 10) =$

Name ______________________________ (7.EE.A.1, HSA-APR.A.1)

Adding and Subtracting Polynomials

Add or subtract the polynomials by combining like terms.

1. $(x^3 - 7x^4 + 2x^2 - 3x) + (4x^4 - 8x^3 + 5x) =$

2. $(6x^2 + 9x + 4) - (^-7x^2 + 2x - 1) =$

3. $(^-8x^2 + 2) + (7x - 5) =$

4. $(8x + 6) - (10x + 16) + (4x - 6) =$

5. $(6x^2 + 7x - 9) + (3x + 8) =$

6. $(4x^2 + 2) + (3x^2 - 4x + 6) - (5x^2 + 10) =$

7. $(2x^2 + 3x + 2) - (6x^3 - 3x^2 + 8) - (^-2x^3 + 9x^2 + 7) =$

8. $(^-3x^2 - 4x^3 - 1) - (2x^3 - 7x - 9) - (2x^3 - 2x^2 - 3) =$

9. $(^-6x^2 - 3x^3 + 4) + (^-7x^3 + 2x + 4) - (^-3x^3 + 5x^2 + 2) =$

10. $(4x^2 + 6x + 3) + (3x^2 - 3x - 2) + (^-4x^2 + 3x - 9) =$

11. $(9x^2 - 7x + {}^-4) + (3x^3 - 4x + {}^-5) + (^-4x^2 - 2x - 5) =$

12. $(3x^3 - 4x + 3) + (^-3x^3 + 5x - 8) =$

13. $(^-5x^5 + 7x^3 - 4x + 2) + (7x^5 - 8x^4 + 9x^3) =$

14. $(4x^2 - 8x + 2) - (^-5x^2 + 5x + 6) =$

15. $(10x^3 - 8x^2 + 6x - 1) - (^-7x^3 + 2x - 4) =$

Name ____________________ 6.EE.A.3, 7.EE.A.1, HSA-APR.A.1

Multiplying Polynomials and Monomials

To multiply a multi-termed polynomial by a monomial, simply multiply each term in the polynomial by the monomial using the distributive property. Remember the properties of exponents when simplifying such problems.

Multiply.	$3x(3x^3 - 2x^2 + 8x - 7)$	
	$3x(3x^3) - 3x(2x^2) + 3x(8x) - 3x(7)$	Distribute 3x to each term.
	$9x^4 - 6x^3 + 24x^2 - 21x$	Multiply to simplify.

You may have to simplify an expression using the distributive property to multiply the polynomial by a monomial.

Simplify.	$4x^7 + 9x^4 - 7 + 2x^2(8x^5 - 5x^2)$	
	$4x^7 + 9x^4 - 7 + 2x^2(8x^5) - 2x^2(5x^2)$	Distribute the 2x.
	$4x^7 + 9x^4 - 7 + 16x^7 - 10x^4$	Multiply to begin simplifying.
	$20x^7 - x^4 - 7$	Simplify.

Multiply each expression.

1. $6x(4x - 3)$

2. $4x(x^2 - 6x + 3)$

3. $(^-3x^2 + 4x - 2)(^-4x^3)$

4. $x(^-5x + 2)$

5. $^-3x(5x^2 - 2x - 6)$

6. $(^-x^2 + 2x - 1)(5x^3)$

7. $^-2x(^-4x - 8)$

8. $^-x^2(7x^2 - x + 4)$

9. $^-5x^4(2x^2 - 8x + 6)$

10. $5y(y^2 - 3y + 1)$

11. $(3x^2 - 4x)(^-x)$

12. $7x(x^3 - x^2)$

Name ______________________________ 6.EE.A.3, 7.EE.A.1, HSA-APR.A.1

Multiplying Polynomials and Monomials

$$4y(y - 3) = 4y^2 - 12y$$

Use the distributive property to multiply the polynomials.

1. $a(a + 8) =$

2. $5b(4b^3 - 6b^2 - 6) =$

3. $3x(x - 3) =$

4. $4a(2a + 6) =$

5. $y(y - 7) =$

6. $^-2x^2(5 - 3x + 3x^2 + 4x^3) =$

7. $4b(3 - b) =$

8. $2xy(2x - 3y) =$

9. $^-5y^2(7y - 8y^2) =$

10. $4x^2(3x^2 - x) =$

11. $x(x^2 + x + x) =$

12. $3b(4b^3 - 12b^2 - 7) =$

13. $(^-7x^3)(3x^2 - 1) =$

14. $^-5ab(6a - 4b) =$

15. $3x(x - 3) =$

16. $^-3x^2(4x^2 - 3x + 3) =$

17. $^-4x^2(3x^3 + 8x^2 + {}^-9x) =$

18. $(3x^4 - 5x^2 - 4)(^-3x^3) =$

Name ______________________________ 6.EE.A.3, 7.EE.A.1, HSA-APR.A.1

Multiplying Polynomials and Monomials

Simplify each expression.

1. $^{-}4x^2 - 5x + 7 + 3(x^2 + 8x - 2)$

2. $5x^2 - 3x(x - 7)$

3. $6x^2 + 4x + (7x - 3)2x$

4. $^{-}x^2 + 8x - 6 - 7(3x^2 - 5x + 9)$

5. $6x^2 + 3(x - 5) - 8x$

6. $2x^2 + 7x + 6 - 5x(^{-}2x - 1)$

7. $^{-}3x^2 - 3x(4x^2 - 5x + 7) - 8x^2$

8. $x^3 + (7x^2 - 9x - 1)x + 10x^2$

Write and simplify an expression to find the area of each polygon. Remember that the area of a triangle can be found using $A = \frac{1}{2}bh$ and the area of a rectangle can be found using $A = lw$.

9.

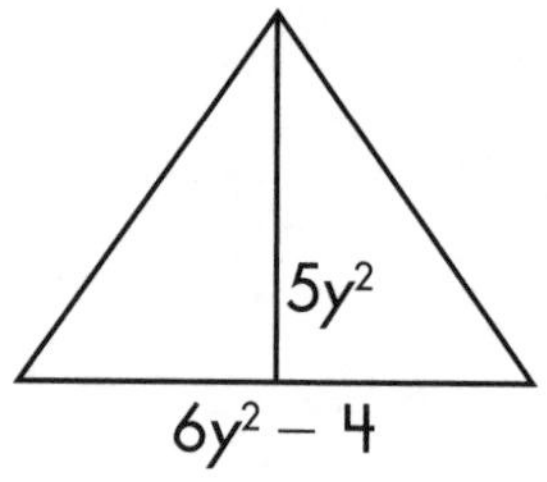

A = ______________________

10.

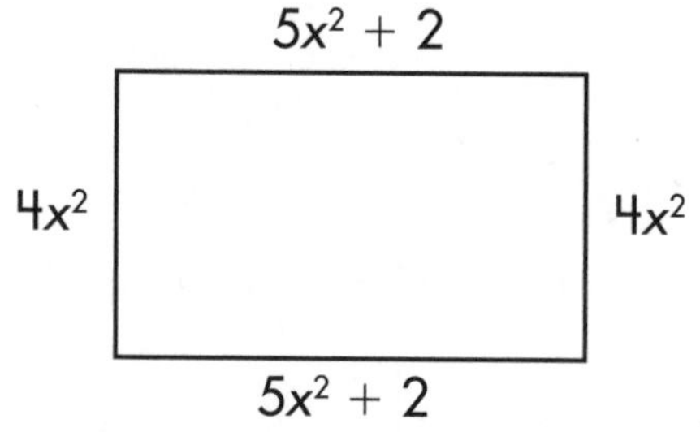

A = ______________________

11.

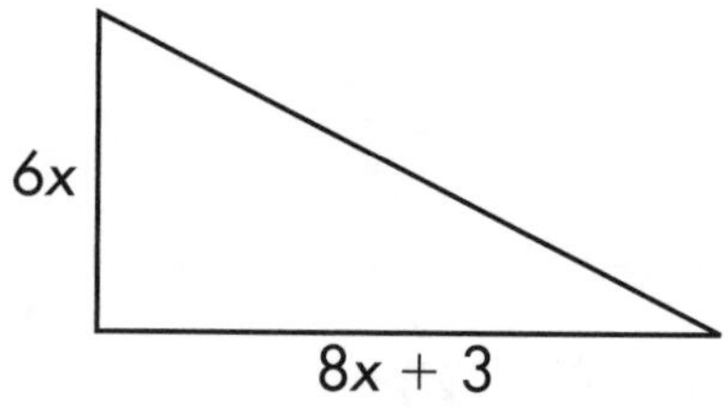

A = ______________________

12.

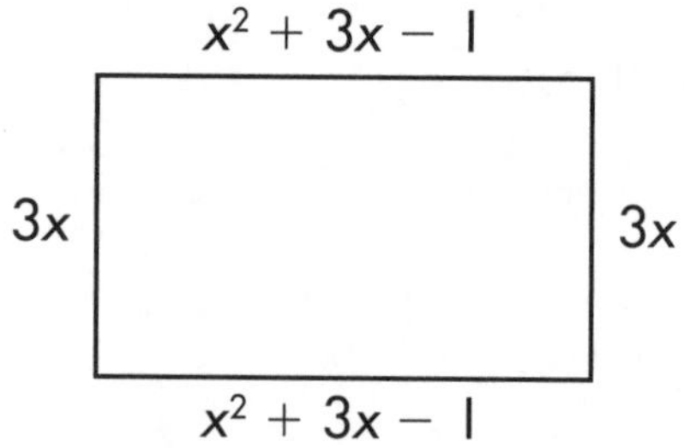

A = ______________________

Name ________________________ 7.EE.A.1, HSA-SSE.A.2, HSA-APR.A.1, HSA-APR.C.4

Multiplying Binomials

The easiest way to multiply binomials is to use what is called the FOIL method. This method multiplies the first terms of the binomials (F), multiplies the outer terms of the binomials (O), multiplies the inner terms of the binomials (I), multiplies the last terms of the binomials (L), and simplifies by adding like terms.

1. Multiply $(3x - 2)(4x + 3)$

$3x(4x) = 12x^2$	Multiply first terms.	F
$3x(3) = 9x$	Multiply outer terms.	O
$^-2(4x) = {}^-8x$	Multiply inner terms.	I
$^-2(3) = {}^-6$	Multiply last terms.	L
$12x^2 + 9x - 8x - 6$	Simplify by combining like terms.	
$12x^2 + x - 6$	Final product	

2. Multiply $(5x + 2)(7x - 3)$

$5x(7x) + 5x(^-3) + 2(7x) + 2(^-3)$
F O I L

$35x^2 - 15x + 14x - 6$	Multiply.
$35x^2 - x - 6$	Final product

Multiply.

1. $(x + 7)(x - 5)$
2. $(x - 5)(3x + 7)$
3. $(9x - 1)(6x + 2)$
4. $(7x - 8)(8x - 7)$
5. $(x + 2)(x + 3)$
6. $(2b - 8)(3b - 7)$
7. $(3x + 5)(^-4x - 7)$
8. $(^-3x - 9)(^-x - 6)$
9. $(x - 10)(x + 1)$
10. $(2x - 3)(5x + 4)$
11. $(^-5x + 6)(x - 2)$
12. $(^-4x + 4)(5x + 8)$
13. $(5b - 2)(3b + 2)$
14. $(3x + y)(3x - 2y)$
15. $(4a + 1)(3a - 3)$
16. $(x - y)(2x + 2y)$

Name ______________________ 7.EE.A.1, HSA-SSE.A.2, HSA-APR.A.1, HSA-APR.C.4

Multiplying Binomials

Rules: $(a + b)(a - b) = a^2 - b^2$
$(a + b)^2 = a^2 + 2ab + b^2$
$(a - b)^2 = a^2 - 2ab + b^2$

Multiply.

1. $(2x + y)(2x - y) =$
2. $(b - 5)(b + 5) =$
3. $(x - y)(2x + 2y) =$
4. $(x - 4y)^2 =$
5. $(7x - 3y)(7x + 3y) =$
6. $(x + 6)(x - 6) =$
7. $(7x + y)(7x - y) =$
8. $(2x - 6y)^2 =$
9. $(3a - b)(3a + b) =$
10. $(x + 2)(x - 2) =$
11. $(12b - 5)(12b + 5) =$
12. $(2x - 3y)^2 =$
13. $(7x - 5y)^2 =$
14. $(3x + 13)(3x - 2) =$
15. $(c + 2d)(c - 2d) =$
16. $(3b + 6)(3b - 6) =$
17. $(2b + 5a)^2 =$
18. $(2x + 3v)^2 =$
19. $(12 + b)(12 - b) =$
20. $(5x + 7)^2 =$

Name ______________________ 7.EE.A.1, HSA-SSE.A.2, HSA-APR.A.1, HSA-APR.C.4

Multiplying Binomials

Multiply.

1. $(4b^2 - 4)(4b^2 + 4) =$

2. $(3b - 3c)^2 =$

3. $(x - 2y)^2 =$

4. $({}^-5x^2 + 3)({}^-5x^2 - 3) =$

5. $(7b^2 - 3c)^2 =$

6. $(4c + 9d)^2 =$

7. $(5x^2 - 5y)^2 =$

8. $(x - yz)(x + yz) =$

9. $({}^-4x + 3y)^2 =$

10. $(4m^2 - 2n)^2 =$

11. $(x^2 - 7y)^2 =$

12. $(4x^2 - 4y^2)^2 =$

13. $(8x^2 - 12)(8x^2 + 12) =$

14. $(2b^2 - 2c^2)^2 =$

15. $(4a + b)^2 =$

16. $(x^2 - 8x)(x^2 + 8x) =$

17. $(2x^2 - y^2)(2x^2 + y^2) =$

18. $(3x^2 - x)(3x^2 + x) =$

19. $(3a - 7b)^2 =$

20. $({}^-6x + 3y)^2 =$

21. $(4x^2 - 4y^2)(4x^2 + 4y^2) =$

22. $({}^-2x^3 + 4)(2x^2 + 5) =$

Name ______________________________ 7.EE.A.1, HSA-SSE.A.2

Factoring Polynomials

To factor a polynomial, write the polynomial as a product of other polynomials.
For example, $3x^2 - 6x$ can be written as $3x(x - 2)$.
$3x$ is the Greatest Common Factor (GCF) of $3x^2$ and $6x$.
$3x$ is a Common Monomial Factor of the terms of the binomial.
$x - 2$ is a Binomial Factor of $3x^2 - 6$.

To factor a trinomial in the form $x^2 + bx + c$, write it as the product of two binomials.
For example, $a^2 - 9a + 14$ can be written as $(a - 7)(a - 2)$.

Look for two numbers that are the product of c and whose sum is b.
If c and b are both positive, the factors will both be positive.
If c is positive and b is negative, the factors will both be negative.
If c is negative, the factors will have opposite signs.

Factor.

1. $3x^2 - 12x^3 =$
2. $2x^3 - x^4 =$
3. $3a^5 - a^3 =$
4. $x^5 + 2x^2 =$
5. $24b^2 + 16b =$
6. $5x^3 - 7x^2 =$
7. $2x^3 + 6x^2 =$
8. $x^3 - 5x^2 =$
9. $x^2 + x - 90 =$
10. $y^2 - 13y + 42 =$
11. $x^2 - x - 6 =$
12. $x^2 - 9x + 18 =$
13. $b^2 - 4b - 45 =$
14. $y^2 - 12y + 36 =$

Name ______________________________ 7.EE.A.1, HSA-SSE.A.2

Factoring Polynomials

$$a^2 - 9a + 14 = (a - 7)(a - 2)$$
$$2x^2 - 5x - 12 = (2x + 3)(x - 4)$$

Factor.

1. $p^2 + 13p + 42 =$
2. $c^2 + c - 30 =$
3. $x^2 + 15xy + 44y^2 =$
4. $x^2 - 13x + 12 =$
5. $x^2 - 13x + 30 =$
6. $x^2 - 8x + 16 =$
7. $x^2 - 12xy + 32y^2 =$
8. $x^2 + 14x + 49 =$
9. $n^2 + 6n - 16 =$
10. $x^2 + 12x + 35 =$
11. $c^2 - 10c + 21 =$
12. $x^2 + 6x - 40 =$
13. $2x^2 + 9x + 10 =$
14. $4x^2 - 18x + 20 =$
15. $3x^2 - 10x + 7 =$
16. $3x^2 - 5x - 12 =$
17. $3x^2 - 4x - 32 =$
18. $5x^2 + 25x + 30 =$

Name ______________________________ 7.EE.A.1, HSA-SSE.A.2

Factoring Polynomials

Factor.

1. $x^2 + 4x + 4 =$

2. $2x^2 + 13x + 6 =$

3. $x^2 + 7x + 12 =$

4. $6x^2 - 21x - 12 =$

5. $x^2 - 8x + 15 =$

6. $9x^2 - 9x - 28 =$

7. $x^2 - 15x + 56 =$

8. $15x^2 + 11x - 14 =$

9. $x^2 + 18x + 45 =$

10. $10x^2 - 28x - 6 =$

11. $x^2 + 4x - 32 =$

12. $7x^2 + 17x + 6 =$

13. $x^2 + 2xy - 63y^2 =$

14. $2x^2 - 2x - 40 =$

15. $x^2 - 14x - 72 =$

16. $11x^2 - 122x + 11 =$

17. $x^2 - xy - 2y^2 =$

18. $2x^2 + 7x + 3 =$

19. $x^2 + 16x + 28 =$

20. $12x^2 + 9x - 3 =$

Name ____________________ 7.EE.A.1, HSA-SSE.A.2

Factoring Polynomials—Special Cases

Factoring Trinomials That Are Quadratic in Form

$x^4 - x^2 - 12 = (x^2)^2 - (x^2) - 12 = (x^2 - 4)(x^2 + 3)$

Factoring the Difference of Two Squares

Rule: $a^2 - b^2 = (a + b)(a - b)$ Example: $x^2 - 49 = (x + 7)(x - 7)$

Factoring Perfect Square Trinomials

Rules: $a^2 + 2ab + b^2 = (a + b)^2$ $a^2 - 2ab + b^2 = (a - b)^2$

Examples: $9x^2 + 6x + 1 = (3x + 1)^2$ $x^2 - 6x + 9 = (x - 3)^2$

Factoring the Sum or Difference of Two Cubes

Rules: $x^3 + y^3 = (x + y)(x^2 - xy + y^2)$ $x^3 - y^3 = (x - y)(x^2 + xy + y^2)$

Examples: $x^3 + 8 = (x + 2)(x^2 - 2x + 4)$ $x^3 - 8 = (x - 2)(x^2 + 2x + 4)$

Factor.

1. $2x^2 - 5x - 12 =$
2. $2x^4 + 16x^2 + 30 =$
3. $x^4 - 8x^2 + 15 =$
4. $7x^4 - 11x^2 - 6 =$
5. $a^2 - 4 =$
6. $b^2 - 9 =$
7. $1 - 9x^2 =$
8. $x^2 - 25 =$
9. $x^2 + 14x + 49 =$
10. $x^2 - 2x + 1 =$
11. $c^2 - 6c + 9 =$
12. $x^2 - 4xy + 4y^2 =$
13. $64x^3 + 1 =$
14. $8x^3 + 27 =$

Name ______________________________

7.EE.A.1, HSA-SSE.A.2

Factoring Polynomials—Special Cases

Factor.

1. $2x^4 - 7x^2 - 15 =$

2. $y^4 + 6y^2 - 16 =$

3. $8x^4 - 23x^2 - 3 =$

4. $6a^6 - 5a^3b^3 - 25b^6 =$

5. $3x^4 + 20x^2 + 33 =$

6. $4x^4y^4 - 2x^2y^2 - 56 =$

7. $y^2 - 81 =$

8. $c^2 - 16 =$

9. $a^2 - 49 =$

10. $49x^2 - 16y^4 =$

11. $16x^2 - 121 =$

12. $25 - x^2y^2 =$

13. $a^2 + 12ab + 36b^2 =$

14. $49x^2 + 28x + 4 =$

15. $x^2 + 14x + 49 =$

16. $c^2 - 20c + 100 =$

17. $x^3 - 1000 =$

18. $1 - 64y^3 =$

19. $x^3 + 125 =$

20. $x^3 - 27 =$

Name ______________________________ 7.EE.A.1, HSA-SSE.A.2

Factoring Polynomials—Special Cases

Factor.

1. $y^4 - y^2 - 12 =$

2. $x^4y^4 - 19x^2y^2 + 34 =$

3. $2x^4y^4 - 17x^2y^2 - 30 =$

4. $x^4y^4 - 8x^2y^2 + 12 =$

5. $64 - x^4y^4 =$

6. $y^2 - 64 =$

7. $81x^2 - 4 =$

8. $16 - 81x^2 =$

9. $x^2y^2 - 121 =$

10. $49x^2 - 36 =$

11. $y^2 - 22y + 121 =$

12. $25a^2 - 40ab + 16b^2 =$

13. $9x^2 - 12x + 4 =$

14. $16x^2 - 40x + 25 =$

15. $9x^2 + 12x + 4 =$

16. $x^2 - 14x + 49 =$

17. $27x^3 - 64 =$

18. $27x^3 - 27 =$

19. $125a^3 - 8b^3 =$

20. $64x^3 - y^3 =$

21. $125x^3 - 64y^3 =$

22. $64x^3 + 27 =$

Name ____________________ HSA-SSE.B.3a, HSA-REI.B.4b

Solving Equations by Factoring

The **Multiplication Property of Zero**: The product of a number and zero is zero.

The **Principle of Zero Products**: If the product of two factors is zero, then at least one of the factors must be zero. This principle is used in solving equations.

Solve: $(x - 4)(x - 5) = 0$

$x - 4 = 0 \quad x - 5 = 0$

$x = 4 \quad x = 5$

The solutions are 4 and 5.

If $(x - 4)(x - 5) = 0$, then $(x - 4) = 0$ or $(x - 5) = 0$.

$x = 4$

$\underline{(4 - 4)(4 - 5) = 0}$

$(0)(^-1) = 0$

$0 = 0$

$x = 5$

$\underline{(5 - 4)(5 - 5) = 0}$

$(1)(0) = 0$

$0 = 0$

Write the solutions for each variable.

1. $x(x + 6) = 0$
2. $b^2 - 81 = 0$
3. $(27 - y)(y - 2) = 0$
4. $z^2 - 1 = 0$
5. $(y - 4)(y - 8) = 0$
6. $y(y - 11) = 0$
7. $8b^2 - 32 = 0$
8. $x^2 - x - 6 = 0$
9. $x^2 - 4x - 21 = 0$
10. $m^2 - 144 = 0$
11. $(y + 5)(y + 6) = 0$
12. $(2x + 4)(x + 7) = 0$
13. $z^2 - 9 = 0$
14. $10x^2 - 10x = 0$
15. $2x^2 - 6x = x - 3$
16. $4y(3y - 2) = 0$
17. $(4y - 1)(y + 2) = 0$
18. $x^2 - 5x + 6 = 0$

Name ______________________________ HSA-SSE.B.3a, HSA-REI.B.4b

Solving Equations by Factoring

$x^2 - 8x = {}^-16$	$x - 4 = 0$
$x^2 - 8x + 16 = 0$	$x = 4$
$(x - 4)(x - 4) = 0$	The solution is 4.

Solve by factoring.

1. $y^2 + 9y = 0$

2. $x - 16 = x(x - 7)$

3. $x - 6 = x(x - 4)$

4. $x^2 + 7x = 0$

5. $x^2 - 4x = 0$

6. $x + 8 = x(x + 3)$

7. $y^2 - y - 6 = 0$

8. $a^2 - 36 = 0$

9. $y^2 + 15 = 8y$

10. $a^2 - 7a = {}^-12$

11. $y^2 + 36y = 0$

12. $3u^2 - 12u - 15 = 0$

13. $y^2 - 8y + 12 = 0$

14. $5a^2 + 25a = 0$

15. $6x^2 + 18x = 0$

16. $2x^2 + x = 6$

17. $x^2 - 5x - 6 = 0$

18. $4x^2 + 16x = 0$

19. $3x^2 - 9x = 0$

20. $y^2 + 5y - 6 = 0$

Name ________________________________ HSA-SSE.B.3a, HSA-REI.B.4b

Solving Equations by Factoring

Solve by factoring.

1. $a^2 - 8a = 0$
2. $x^2 = 3x + 4$
3. $4a^2 + 15a - 4 = 0$
4. $x^2 - x - 6 = 0$
5. $3x^2 - 13x + 4 = 0$
6. $6x^2 = 23x + 18$
7. $x^2 + 7x + 12 = 0$
8. $x^2 + 5x - 6 = 0$
9. $x^2 = 6x + 7$
10. $x^2 = 10x - 25$
11. $x^2 + 3x - 10 = 0$
12. $x^2 - 6x + 9 = 0$
13. $y^2 - 3y + 2 = 0$
14. $2x^2 - 9x + 9 = 0$
15. $r^2 - 15r = 16$
16. $x^2 + 7x + 10 = 0$
17. $3x^2 - 2x - 8 = 0$
18. $2a^2 + 4a - 6 = 0$
19. $x^2 + 3x - 4 = 0$
20. $4a^2 + 9a + 2 = 0$
21. $9x^2 = 18x + 0$
22. $2x^2 = 9x + 5$

Name ______________________________ 7.EE.B.4, HSA-SSE.B.3a, HSA-REI.B.4b

Writing and Solving Quadratic Equations

Problem solving often involves quadratic equations.

The square of a number is 8 less than 6 times the number. Find the numbers that make the sentence true.

$x^2 = 6x - 8$	First, set up the equation.
$x^2 - 6x + 8 = 0$	Put the equation in standard form.
$(x - 2)(x - 4) = 0$	Factor.
$x - 2 = 0$ or $x - 4 = 0$	Solve for x.
$x = 2$ or $x = 4$	So, the numbers are 2 and 4.

The product of a number and 10 more than 4 times the number is 50. Find the numbers that make the sentence true.

$x(4x + 10) = 50$	First, set up the equation.
$4x^2 + 10x - 50 = 0$	Put the equation in standard form.
$2(2x - 5)(x + 5) = 0$	Factor.
Since $2 \neq 0$, then $2x - 5 = 0$ or $x + 5 = 0$	Solve for x.
$x = \frac{5}{2}$ and $x = {}^{-}5$	So, the numbers are $\frac{5}{2}$ and $^{-}5$.

Write an equation for each word problem and solve it.

1. Six less than 3 times a number is 12 less than twice the number.

   Equation ______________________ **Solution** ______________________

2. The square of a number is 8 more than twice the number.

   Equation ______________________ **Solution** ______________________

3. The sum of the square of a number and 4 times the number is 12.

   Equation ______________________ **Solution** ______________________

4. Twice the square of a number is 4 more than twice the number.

   Equation ______________________ **Solution** ______________________

5. Three times the square of a number is 5 less than 16 times the number.

   Equation ______________________ **Solution** ______________________

Name ______________________________ (7.EE.B.4, HSA-SSE.B.3a, HSA-REI.B.4b)

Writing and Solving Quadratic Equations

The length of a rectangle is 3 inches longer than the width. The area of the rectangle is 40 square inches. Find the length and width of the rectangle.

Width of rectangle: w
Length of rectangle: $w + 3$
$A = lw$
$40 = (w + 3)(w)$
$40 = w^2 + 3w$
$0 = w^2 + 3w - 40$
$0 = (w + 8)(w - 5)$
$w + 8 = 0$ or $w - 5 = 0$
$w = {}^-8 \quad w = 5$

Since the width cannot be a negative number, the width is 5.

$l = 5 + 3 = 8$

The length is 8 inches, and the width is 5 inches.

Write an equation for each word problem and solve it.

1. The area of a square is 121 m^2. Find the length of the sides of the square.

Equation ______________________ **Solution** ______________________

2. The area of a rectangle is 72 m^2. Its length is twice its width. Find the length and width of the rectangle.

Equation ______________________ **Solution** ______________________

3. The area of a rectangle is 36 cm^2. Its width is 4 times its length. Find the length and width of the rectangle.

Equation ______________________ **Solution** ______________________

4. The width of a rectangle is 5 more than twice its length. The area of the rectangle is 33 $in.^2$. Find the dimensions of the rectangle.

Equation ______________________ **Solution** ______________________

5. The length of a rectangle is 4 more than twice its width. The area of the rectangle is 96 $ft.^2$. Find its dimensions.

Equation ______________________ **Solution** ______________________

Name ______________________________ 7.EE.B.4, HSA-SSE.B.3a, HSA-REI.B.4b

Writing and Solving Quadratic Equations

Write an equation for each word problem and solve it.

1. The length of a rectangle is 5 inches longer than the width. The area of the rectangle is 50 square inches. Find the length and width.

   Equation ______________________ **Solution** ______________________

2. The length of a rectangle is 2 more than 8 times its width. Its area is 36 cm^2. Find the width and length.

   Equation ______________________ **Solution** ______________________

3. The sum of a number and its square is 42. Find the numbers.

   Equation ______________________ **Solution** ______________________

4. The sum of a number and its square is 56. Find the numbers.

   Equation ______________________ **Solution** ______________________

5. The length of a rectangle is twice its width. Its area is 32 m^2. Find the length and width.

   Equation ______________________ **Solution** ______________________

6. The square of a number is 80 more than 2 times the number. Find the numbers.

   Equation ______________________ **Solution** ______________________

7. The square of a number is 48 more than 2 times the number. Find the numbers.

   Equation ______________________ **Solution** ______________________

8. The width of a rectangle is 2/3 the length. The area is 54 in^2. Find the length and width.

   Equation ______________________ **Solution** ______________________

Name ______________________ HSA-APR.D.6

Dividing Polynomials

Simplify using long division: $(x^2 + 6x + 5) \div (x + 1)$

$$\frac{(x^2 + 6x + 5)}{(x + 1)}$$

$$\begin{array}{r} x + 5 \\ x + 1\overline{)x^2 + 6x + 5} \\ \underline{-\ x^2 + 1x} \\ 5x + 5 \\ \underline{-\ 5x + 5} \\ 0 \end{array}$$

Divide by using long division.

1. $(x^2 + 5x + 6) \div (x + 3) =$

2. $(x^2 + 4x - 21) \div (x - 3) =$

3. $(x^2 - 3x - 40) \div (x + 5) =$

4. $(x^2 - x - 42) \div (x + 6) =$

5. $(x^2 - 8x + 16) \div (x - 4) =$

6. $(x^2 + 2x - 35) \div (x + 7) =$

7. $(x^2 - 6x + 9) \div (x - 3) =$

8. $(x^2 + 5x + 4) \div (x + 1) =$

9. $(x^2 + 7x + 10) \div (x + 2) =$

10. $(x^2 + 9x + 8) \div (x + 8) =$

Name ______________________________

Dividing Polynomials

Simplify using synthetic division: $(2x^3 + 3x^2 - 4x + 8) \div (x + 3)$

$$\begin{array}{r|rrrr} {}^{-}3 & 2 & 3 & {}^{-}4 & 8 \\ & & {}^{-}6 & 9 & {}^{-}15 \\ \hline & 2 & {}^{-}3 & 5 & {}^{-}7 \end{array}$$

$$= 2x^2 - 3x + 5 - \frac{7}{x + 3}$$

Divide by using synthetic division.

1. $(x^2 + x - 2) \div (x + 2) =$

2. $(5x^2 - 12x - 9) \div (x - 3) =$

3. $(3x^2 - 5) \div (x - 1) =$

4. $(3x^3 + 8x^2 + 9x + 10) \div (x + 2) =$

5. $(x^3 - 4x^2 - 36x - 16) \div (x + 4) =$

6. $(3x^2 - 7x + 6) \div (x - 3) =$

7. $(4x^2 + 9x + 6) \div (x + 1) =$

8. $(4x^2 + 23x + 28) \div (x + 4) =$

9. $(3x^2 + 19x + 20) \div (x + 5) =$

10. $(x^2 + 14x + 45) \div (x + 5) =$

Name ______________________________ HSA-APR.D.6

Dividing Polynomials

Divide.

1. $\frac{x^2 + 9x + 8}{x - 1} =$

2. $\frac{2x^2 - 5x - 3}{x - 3} =$

3. $\frac{4x^2 - 7x - 2}{4x + 1} =$

4. $\frac{2x^2 + x - 3}{x - 1} =$

5. $\frac{21x^2 + 22x - 8}{3x + 4} =$

6. $\frac{3x^2 - 27}{x + 3} =$

7. $\frac{9x^2 - 27x - 36}{x - 4} =$

8. $\frac{5x^2 + 43x - 18}{x + 9} =$

9. $\frac{x^2 - 4x - 45}{x + 5} =$

10. $\frac{2x^2 - x - 21}{x + 3} =$

11. $(3x^3 - 13x^2 - 13x + 15) \div (x - 5) =$

12. $(2x^3 - 12x^2 + 5x - 27) \div (x - 6) =$

13. $(3x^2 - 75) \div (x - 5) =$

14. $(2x^2 + 7x - 10) \div (x + 1) =$

Name ____________________ HSA-APR.D.7

Operations with Rational Expressions

Add or subtract:

$$\frac{7x - 12}{2x^2 + 5x - 12} - \frac{3x - 6}{2x^2 + 5x - 12} = \frac{(7x - 12) - (3x - 6)}{2x^2 + 5x - 12} = \frac{4x - 6}{(2x - 3)(x + 4)} =$$

$$\frac{2\cancel{(2x - 3)}}{\cancel{(2x - 3)}(x + 4)} = \frac{2}{x + 4}$$

Multiply:

$$\frac{4x + 16}{12x + 48} \bullet \frac{(x - 4)}{2x} = \frac{\cancel{4(x + 4)}(x - 4)}{\cancel{24x(x + 4)}} = \frac{x - 4}{6x}$$

Divide:

$$\frac{x + 2}{x + 3} \div \frac{x^2 - 4}{x - 2} = \frac{\cancel{x + 2}}{x + 3} \bullet \frac{\cancel{x - 2}}{\cancel{(x + 2)}\cancel{(x - 2)}} = \frac{1}{x + 3}$$

Simplify.

1. $\frac{x^2 + x - 6}{x + 1} \bullet \frac{x + 1}{x^2 - 9} =$

2. $\frac{x^2 - 1}{x^2 - 2x - 3} \bullet \frac{x + 4}{6x - 6} =$

3. $\frac{x - 7}{x + 2} \div \frac{x^2 - 49}{x^2 + 9x + 14} =$

4. $\frac{x + 2}{4x(x - 6)} \div \frac{x^2 - 4}{8x(x - 6)} =$

5. $\frac{2}{x + 2} + \frac{6x}{x^2 - 4} =$

6. $\frac{2}{4xy} + \frac{14}{3xy} - \frac{9}{2xy} =$

Name ______________________________ HSA-APR.D.7

Operations with Rational Expressions

Simplify.

1. $\frac{x + 2}{x - 3} \bullet \frac{x^2 - 8x + 15}{5x - 25} =$

2. $\frac{12x^2y^4}{36ab^3} \bullet \frac{6a^2v^3}{48xy^4} =$

3. $\frac{x^2 + 6x + 8}{x^2 - 16} \bullet \frac{3x - 12}{4x + 4} =$

4. $\frac{x^2 - 2x - 8}{x^2 - 4} \bullet \frac{x - 2}{x + 3} =$

5. $\frac{x^2 + 3x + 2}{x + 7} \bullet \frac{x^2 + 9x + 14}{x^2 + 4x + 4} =$

6. $\frac{2x^2 + 12x + 18}{x^2 + 5x - 6} \div \frac{2x + 6}{x - 1} =$

7. $\frac{2x^2 + 6x}{x^2 + 2x} \div \frac{x^2 - 9}{4x - 12} =$

8. $\frac{15x^4y^2}{5xy} \div \frac{10x^3y}{5y^2} =$

9. $\frac{x^2 + 8x}{x^2 + 14x + 48} \div \frac{x^2 + x}{x^2} =$

10. $\frac{3x^2 + 6x}{x^2 + 6x} \div \frac{x^2 - 4}{2x - 4} =$

11. $\frac{x}{x - 4} + \frac{4}{x^2 - x - 12} =$

12. $\frac{4}{3x - 8} - \frac{x}{4x - 7} =$

13. $-\frac{4}{2x^2} + \frac{5}{2x^2} + \frac{8}{3x^2} =$

14. $\frac{4x}{x^2 + x - 2} - \frac{4x}{x^2 + x - 2} =$

Name ____________________ HSA-APR.D.7

Operations with Rational Expressions

Simplify.

1. $\frac{2x^2 - 32}{2x + 8} \bullet \frac{x^2 - 9}{x^2 - 3x - 4} =$

2. $\frac{14x^2y^4}{42a^2b^4} \bullet \frac{28a^2b^3}{35x^3y^4} =$

3. $\frac{x^2 - 100}{x - 5} \bullet \frac{x + 5}{x^2 - 5x - 50} =$

4. $\frac{x^2 - 12x + 35}{x^2 - 5x - 14} \bullet \frac{x^2 + 7x + 10}{x^2 - 25} =$

5. $\frac{9x^2 - 25}{4x^2 + 4x - 3} \bullet \frac{40x^2 - 10x}{3x + 5} =$

6. $\frac{x + 4}{6x^2 + 24} \bullet \frac{2x^3 - 8x}{x^2 + 4x} =$

7. $\frac{x^2 - 7x}{x^2 - 14x + 49} \div \frac{2x^2 + 6x}{x^2 + x - 56} =$

8. $\frac{x^2 - 16}{x^2 + 7x + 12} \div \frac{5x - 20}{x + 3} =$

9. $\frac{27x^6y^2}{9x^3y} \div \frac{4x^5y^3}{16x^4y^2} =$

10. $\frac{x^2 + 9x - 10}{x^2 + 5x - 14} \div \frac{3x + 30}{2x - 4} =$

11. $\frac{24a^4b^2}{8a^2b} \div \frac{12ab^3}{16a^3b} =$

12. $\frac{11x}{x^2 - 6x - 7} + \frac{5x}{x^2 + 9x + 8} =$

13. $\frac{3x}{x^2 - 4} + \frac{5x}{x^2 - 4} =$

14. $-\frac{10}{4x^2} + \frac{6}{4x^2} + \frac{8}{4x^2} =$

15. $\frac{5}{2xy} + \frac{5}{4xy} - \frac{12}{6xy} =$

16. $\frac{8}{4x + 16} - \frac{1}{x + 4} - \frac{3}{4} =$

Name ______________________________ 7.RP.A.1

Ratios and Proportions

Solve the following ratio for x.

$\frac{x}{5} = \frac{4}{10}$ → Take cross products and solve. → $\frac{x}{5} \times \frac{4}{10}$ → $5 \bullet 4 = 20$, $x \bullet 10 = 10x$

→ $10x = 20$ → $\frac{10x}{10} = \frac{20}{10}$ → $x = 2$

Solve.

1. $\frac{4}{(x - 3)} = \frac{28}{49}$

2. $\frac{(5 + x)}{10} = \frac{2}{5}$

3. $\frac{x}{30} = \frac{7}{10}$

4. $\frac{(x - 2)}{16} = \frac{x}{4}$

5. $\frac{2}{x} = \frac{6}{30}$

6. $\frac{(x + 1)}{7} = \frac{6}{14}$

7. $\frac{x}{15} = \frac{5}{75}$

8. $\frac{x}{20} = \frac{2}{10}$

9. $\frac{x}{6} = \frac{(x - 3)}{12}$

10. $\frac{x}{5} = \frac{12}{6}$

11. $\frac{6}{(x + 5)} = \frac{18}{24}$

12. $\frac{5}{15} = \frac{x}{9}$

13. $\frac{+10}{\quad} = \frac{5}{2}$

14. $\frac{x}{3} = \frac{12}{27}$

Name ____________________ 6.RP.A.3, 7.RP.A.1, 7.RP.A.2c, 7.RP.A.3

Ratios and Proportions

Three liters of soda cost $3.00. At this rate, how much would 10 liters of soda cost? To find the cost, write and solve a ratio using *x* to represent the cost.

$\frac{\text{liters}}{\text{cost}} \rightarrow \frac{3}{3.00} = \frac{10}{x} \rightarrow 3x = 10(3.00) \rightarrow 3x = 30 \rightarrow \frac{3x}{3} = \frac{30}{3}$

$\rightarrow x = 10 \rightarrow$ The cost of 10 liters of soda is $10.00.

Solve.

1. $\frac{4}{x+8} = \frac{3}{x}$

2. $\frac{x}{4} = \frac{21}{7}$

3. $\frac{9}{3} = \frac{6}{x}$

4. $\frac{x-3}{4} = \frac{x-4}{3}$

5. $\frac{x}{4} = \frac{8}{2x}$

6. $\frac{12}{x} = 3$

Set up a proportion and solve.

7. A copy machine can print 120 pages per minute. At this rate, how many minutes will it take to make 840 copies?

8. Two gallons of fruit juice will serve 35 people. How much fruit juice is needed to serve 105 people?

9. A stock investment of $4,000 earns $360 in interest each year. At the same rate, how much interest will a person earn if he invests $6,000?

Name ______________________ 6.RP.A.3, 7.RP.A.1, 7.RP.A.2c, 7.RP.A.3

Ratios and Proportions

Set up a proportion and solve.

1. An investment of $36,000 earns $900 each year. At the same rate, how much money must be invested to earn $1,200 each year?

2. The sales tax on a $15,000 car is $540. At this rate, what is the tax on a $32,000 car?

3. Six gallons of paint will cover 120 doors. At this rate, how many gallons of paint are needed to cover 480 doors?

4. A lawnmower can cut 1 acre on 0.5 gallons of gasoline. At this rate, how much gasoline is needed to cut 3.5 acres?

5. An aerobics instructor burns 400 calories in 1 hour. How many hours would the instructor have to do aerobics to burn 660 calories?

6. The real estate tax for a house that costs $56,000 is $1,400. At this rate, what is the value of a house for which the real estate tax is $1,800?

7. One hundred thirty-six tiles are required to tile a 36 $ft.^2$ area. At this rate, how many tiles are required to tile a 288 $ft.^2$ area?

Graphing Ordered Pairs

Start at the origin (0, 0).

(x, y) = (1, ⁻2)	right 1 and down 2
(x, y) = (⁻3, 4)	left 3 and up 4
(x, y) = (2, ⁻2)	right 2 and down 2

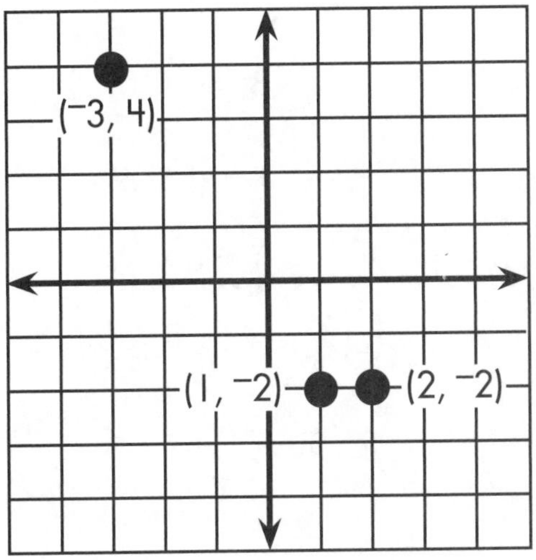

Label the following points. Start at the origin (0, 0).

A (6, 2)

B (6, 3)

C (5, 1)

D (5, 0)

E (1, ⁻3)

F (2, ⁻6)

G (5, 4)

H (2, 4)

I (4, 3)

J (1, ⁻2)

K (5, ⁻3)

L (1, 4)

M (3, 1)

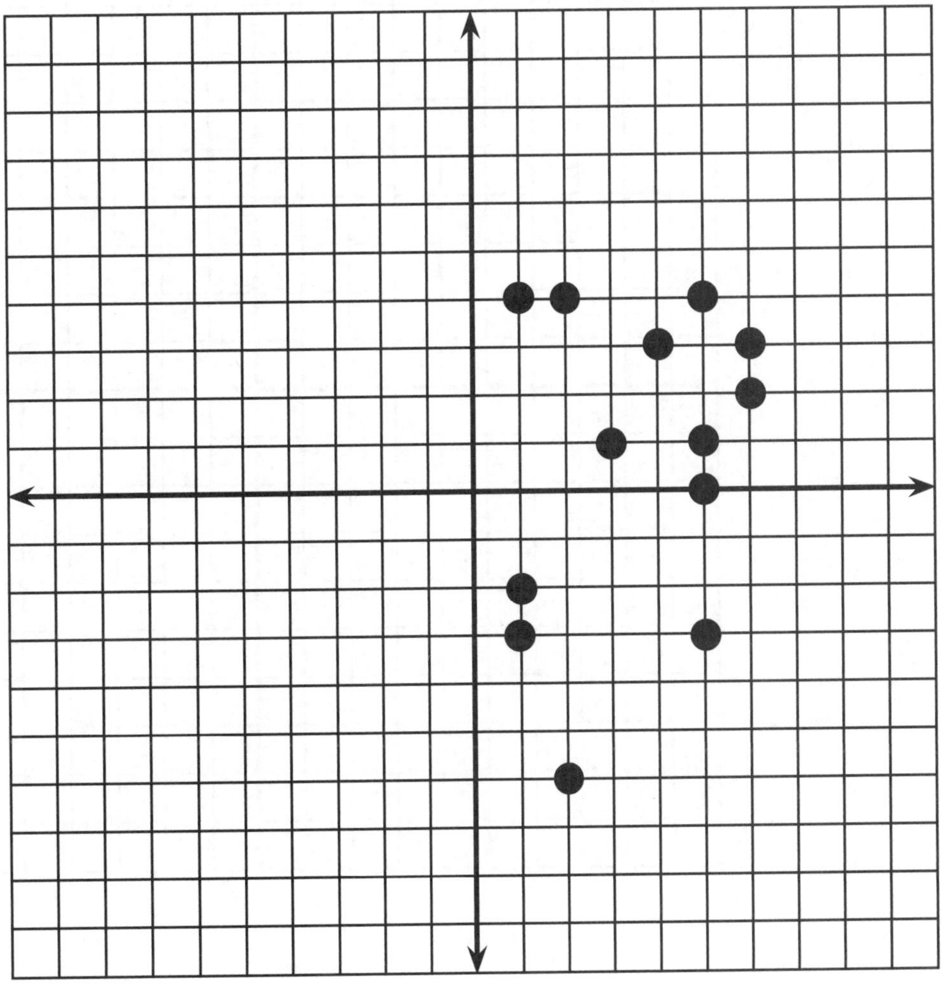

Name ______________________________

5.G.A.1, 6.NS.C.6b, 6.NS.C.6c

Graphing Ordered Pairs

Graph and label the following points.

A $(^-1, 2)$	B $(^-1, 0)$	C $(^-1, ^-6)$	D $(^-4, 2)$
E $(^-6, 2)$	F $(^-5, 6)$	G $(^-4, 1)$	H $(2, ^-8)$
I $(6, 6)$	J $(^-7, 5)$	K $(^-1, ^-1)$	L $(^-3, ^-3)$
M $(6, ^-4)$			

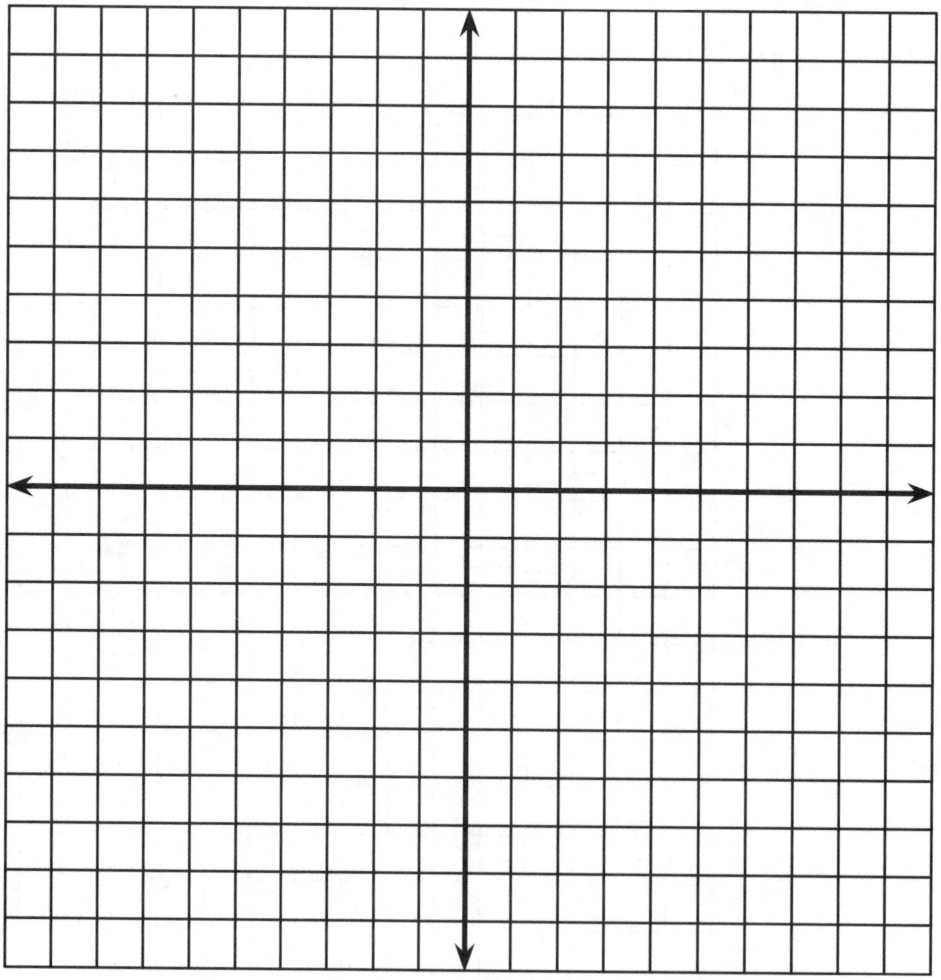

Name ____

6.NS.C.6, 6.EE.B.6, 8.F.A.1, HSA-REI.D.10

Graphing Ordered Pairs

Solve for each *x* value. Graph each ordered pair.

1. $x + y = 4$

x	y
4	
5	
6	

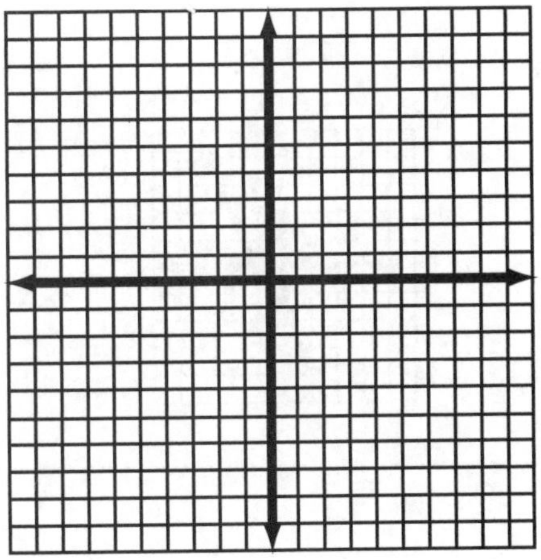

2. $y = 3 - x$

x	y
5	
−2	
0	

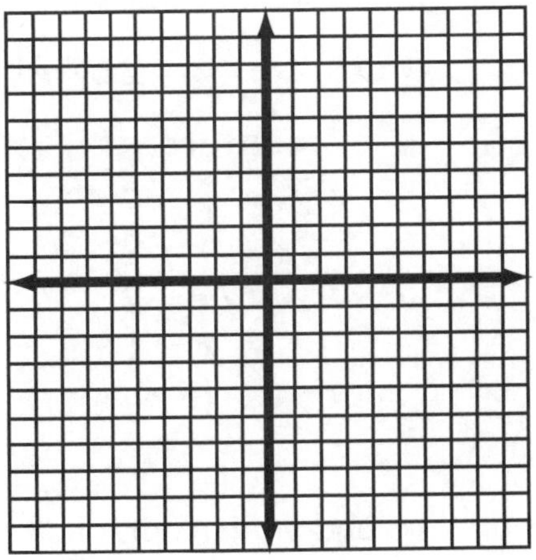

3. $4x + y = 6$

x	y
0	
1	
2	

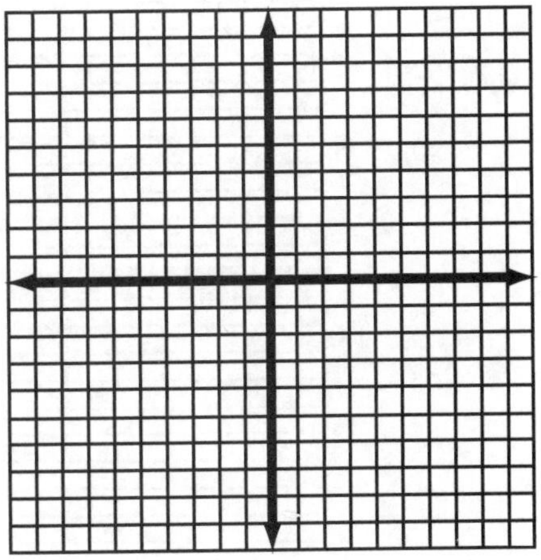

4. $y = 2x - 7$

x	y
3	
4	
5	

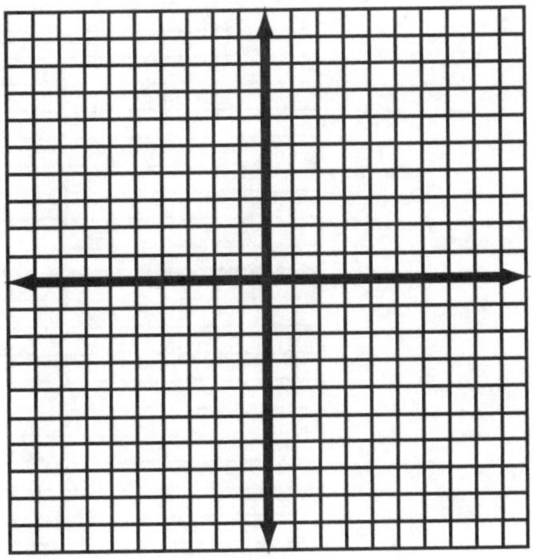

5. $y = x + 3$

x	y
2	
0	
−3	

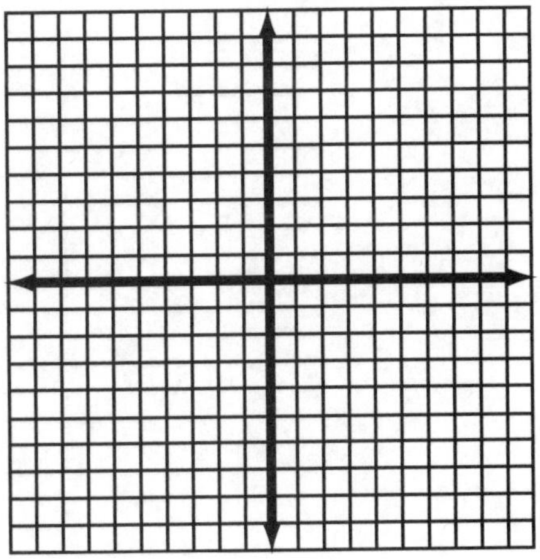

6. $y = 9 - x$

x	y
0	
4	
3	

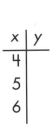

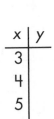

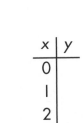

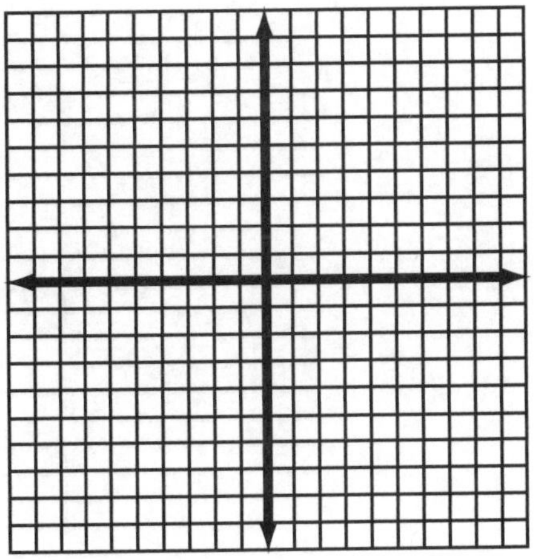

Name ______________________________ 8.EE.B.6, 8.F.A.1

Graphing Linear Equations

One way to graph a linear equation is to find the x- and y-intercepts. To find the x-intercept, let $y = 0$. To find the y-intercept, let $x = 0$. Plot the two points and connect with a straight line.

Graph $2x - y = 2$ by using the x- and y-intercepts.

***x*-intercept**	***y*-intercept**
$2x - y = 2$	$2x - y = 2$
$2x - 0 = 2$	$2(0) - y = 2$
$2x = 2$	$^-y = 2$
$x = 1$	$y = {}^-2$
$(1, 0)$	$(0, {}^-2)$

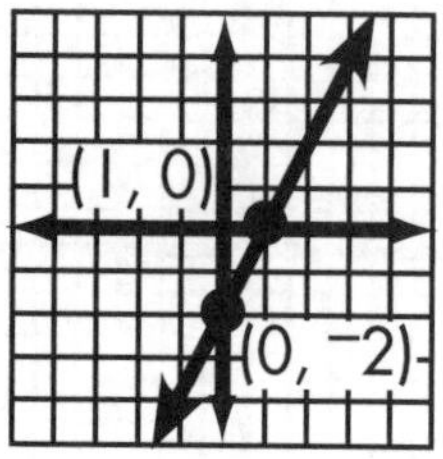

Find the x- and y-intercepts and graph.

1. $3x - 9y = 18$

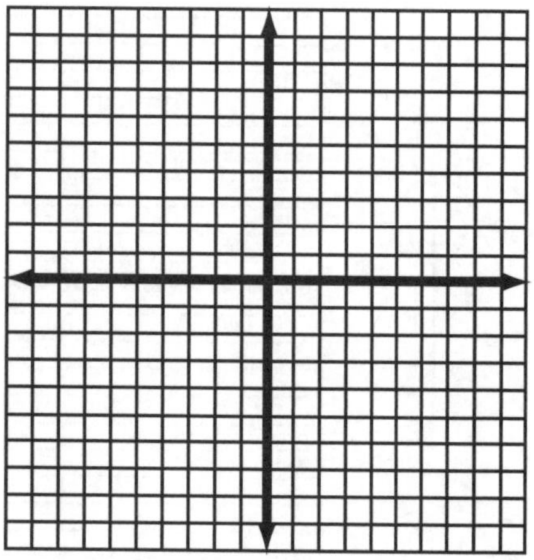

2. $4x + 2y = {}^-8$

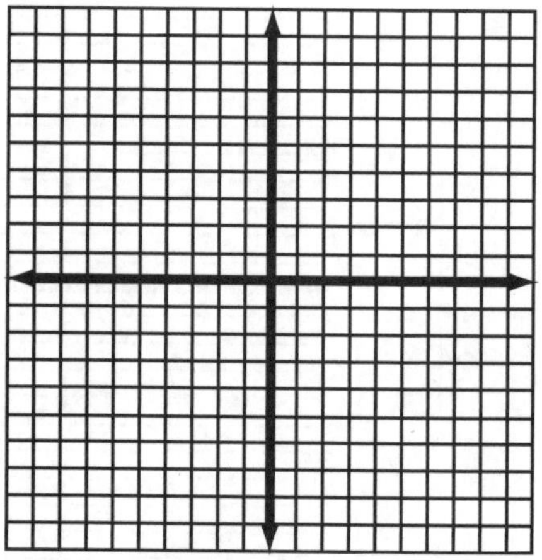

3. $2x + 4y = 20$

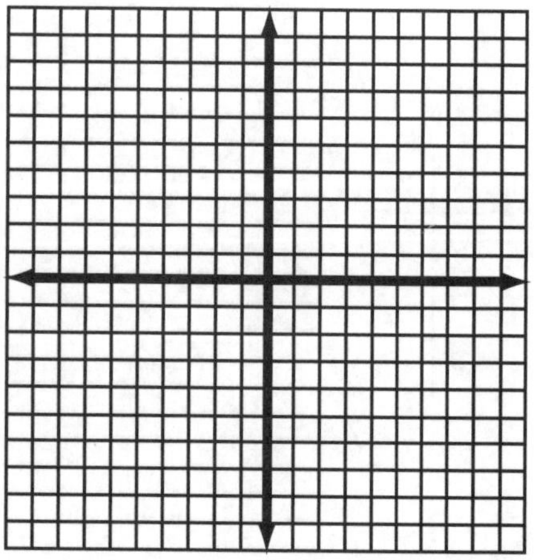

4. $x + 5y = 10$

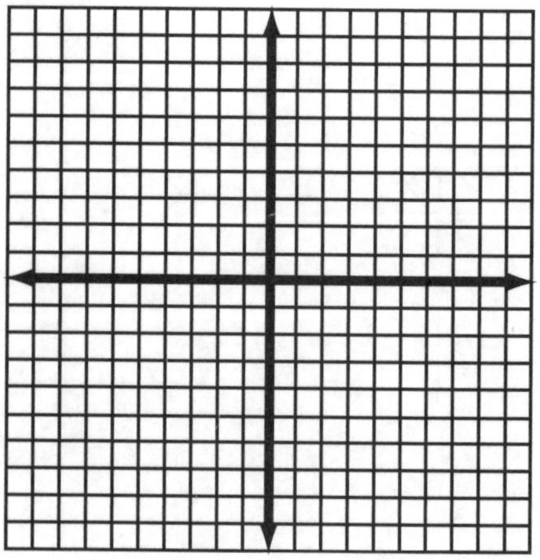

Name ____________________ 8.EE.B.6, 8.F.A.1

Graphing Linear Equations

Another way to graph a linear equation is to use the **slope-intercept form**. A linear equation written in slope-intercept form is $y = mx + b$, where m is the slope of the line and b is the y-intercept.

Example:		
$y = 3x + 2$	$y = {}^{-}5x + 1$	$y = x - 7$
$m = 3$, y-int. $= 2$	$m = {}^{-}5$, y-int. $= 1$	$m = 1$, y-int. $= {}^{-}7$

You can rewrite an equation in slope-intercept form by adding or subtracting terms from both sides and simplifying.

To graph the equation, plot the y-intercept. Starting at the y-intercept, move up or down and then right or left according to the slope and make a new point.

Example: If $m = 4$ and $b = 3$, the y-intercept is (0, 3). The next point to graph would be (1, 7) since the slope is 4, or $\frac{4}{1}$.

Solve for y, state the m and b, and graph.

1. $y = {}^{-}3x + 5$

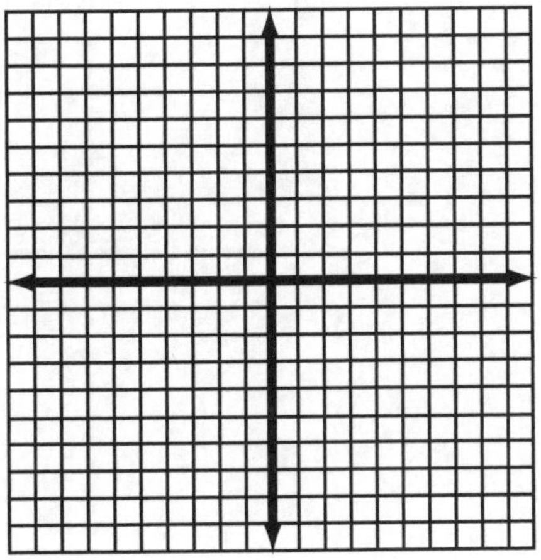

2. ${}^{-}8x + y = {}^{-}4$

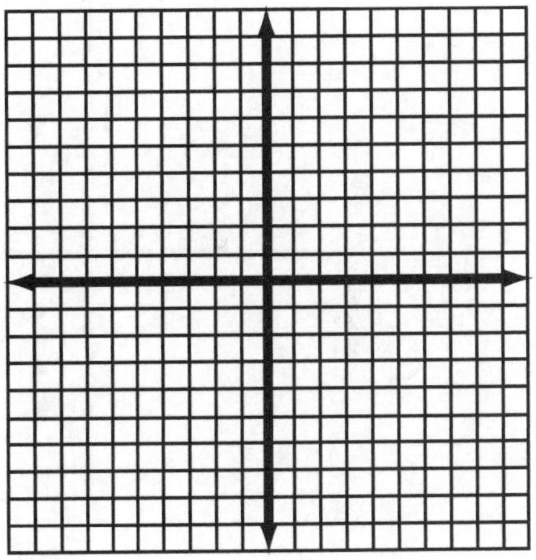

3. $x - y = 7$

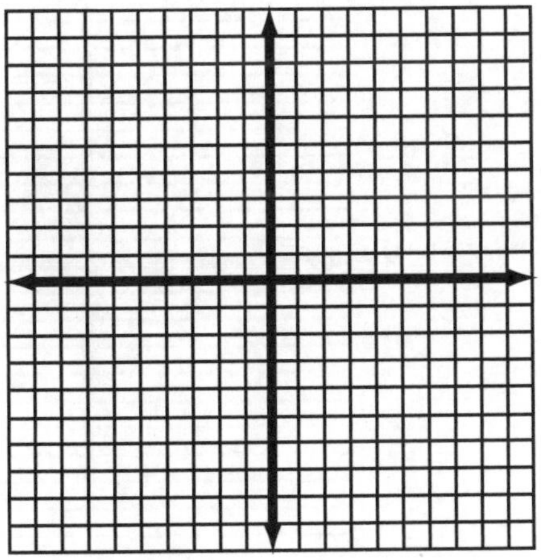

4. $2y = 3x - 4$

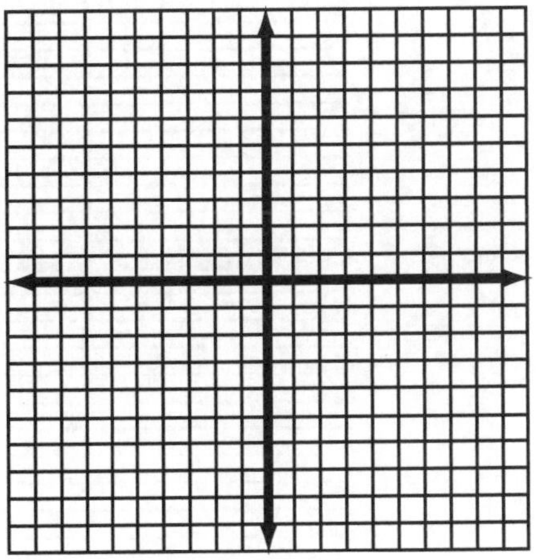

Name ______________________________ 8.EE.B.6, 8.F.A.1

Graphing Linear Equations

Graph each equation using either slope-intercept form or *x*- and *y*-intercepts.

1. $2x + 3y = 9$

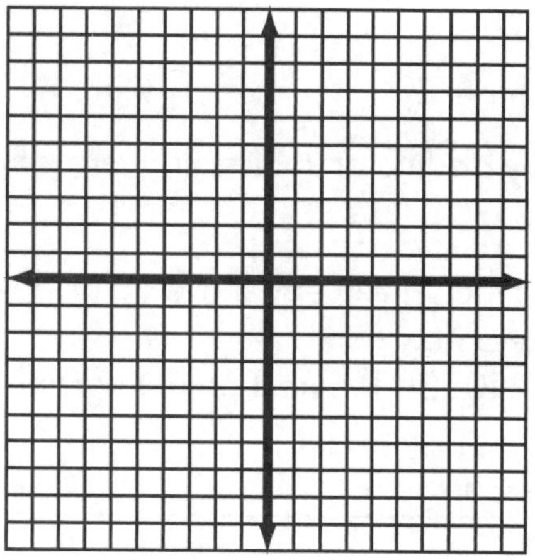

2. $x - 3y + 9 = 0$

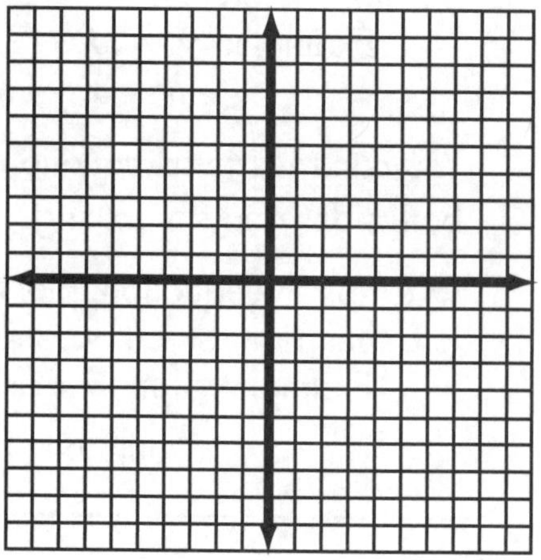

3. $6x + 2y = 12$

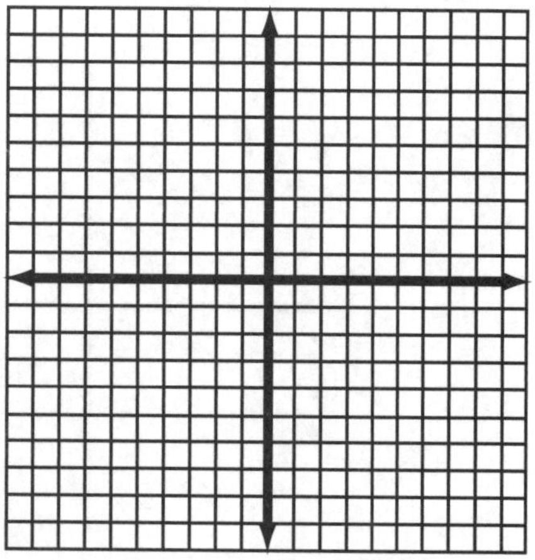

4. $y = 2x - 4$

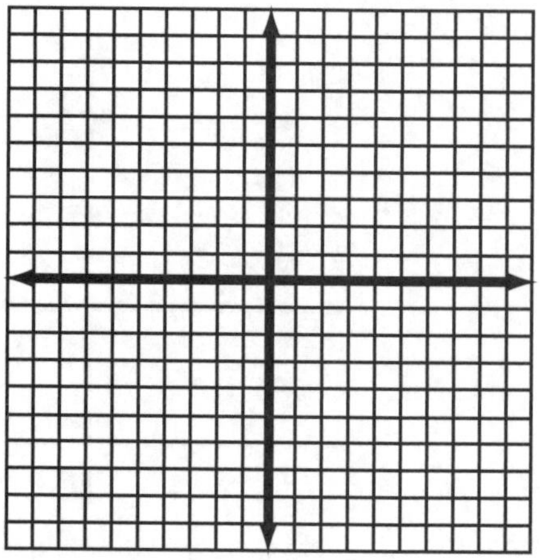

5. $5x - y = 7$

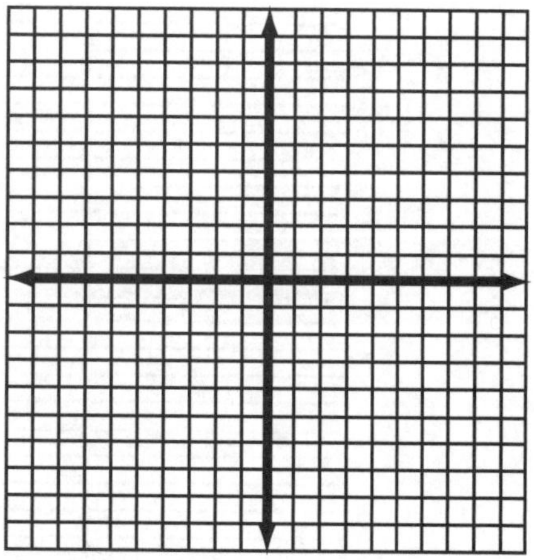

6. $x - 3y + 6 = 0$

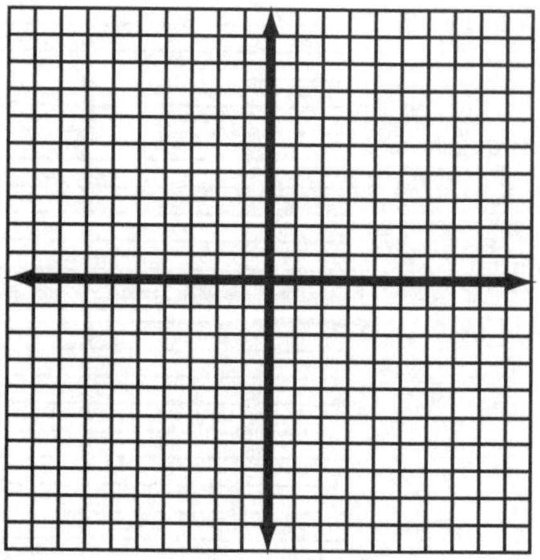

Name ______________________________ 8.EE.B.6, 8.F.A.3

Writing Linear Equations

This equation, $y = mx + b$, is the **slope-intercept of a straight line.**
For all equations of the form $y = mx + b$, m is the slope of the line.
The y-intercept is $(0, b)$.

Find the equation of a line using the slope-intercept form: $y = mx + b$.

1. $m = 4$, y-int. $= {}^-1$

2. $m = {}^-2$, y-int. $= {}^-5$

3. $m = 1$, y-int. $= 2$

4. $m = \frac{7}{5}$ $b = {}^-2$

5. $m = \frac{3}{4}$ $b = \frac{2}{3}$

6. $m = \frac{3}{7}$ $b = \frac{1}{3}$

7. $m = \frac{3}{5}$ $b = \frac{1}{5}$

8. $m = {}^-4$ $b = \frac{3}{4}$

9. $m = 0$ $b = 5$

10. $m = \frac{1}{4}$ $b = \frac{2}{3}$

Name ______________________________ 8.EE.B.6, 8.F.A.3

Writing Linear Equations

Slope-Intercept Formula
$y = mx + b$
m: slope
b: y-intercept containing coordinate points (o, b)

$m = 4$, passing through points $(1, 2)$
Using this information, $\{m = 4, x = 1, y = 2\}$.
Substitute into $y = mx + b$ to find b.

$2 = (4)(1) + b \longrightarrow 2 = 4 + b$

$\longrightarrow 2 - 4 = 4 - 4 + b \longrightarrow {}^{-}2 = b$

With the information $m = 4$, $b - {}^{-}2$, write the equation as $y = 4x - 2$.

Find the equation of the line with the given slope passing through the indicated point (P).

1. $p = ({}^{-}2, {}^{-}6)$, $m = 3$
   $b =$
   Equation ______________________

2. $p = ({}^{-}4, 3)$, $m = 1$
   $b =$
   Equation ______________________

3. $p = (3, 5)$, $m = 0$
   $b =$
   Equation ______________________

4. $p = (5, 7)$, $m = 3$
   $b =$
   Equation ______________________

5. $p = ({}^{-}7, {}^{-}7)$, $m = {}^{-}7$
   $b =$
   Equation ______________________

6. $p = (2, {}^{-}6)$, $m = 4$
   $b =$
   Equation ______________________

7. $p = (2, 4)$, $m = 4$
   $b =$
   Equation ______________________

8. $p = (6, {}^{-}1)$, $m = {}^{-}5$
   $b =$
   Equation ______________________

9. $p = ({}^{-}1, 1)$, $m = 2$
   $b =$
   Equation ______________________

10. $p = ({}^{-}1, {}^{-}6)$, $m = 2$
    $b =$
    Equation ______________________

11. $p = (4, 5)$, $m = {}^{-}2$
    $b =$
    Equation ______________________

12. $p = (2, 6)$, $m = {}^{-}6$
    $b =$
    Equation ______________________

Name ______________________________

8.EE.B.6, 8.F.A.3

Writing Linear Equations

Write the equation of the line that passes through the point and has the given slope. Put each equation in slope-intercept form.

1. $p = (2, 3)$, $m = 4$

2. $p = (^-5, ^-4)$, $m = ^-8$

3. $p = (^-5, 7)$, $m = ^-3$

4. $p = (2, 4)$, $m = ^-1$

5. $p = (^-1, 2)$, $m = 1$

6. $p = (6, ^-8)$, $m = 2$

7. $p = (1, ^-1)$, $m = \frac{1}{2}$

8. $p = (^-4, ^-1)$, $m = -\frac{1}{2}$

9. $p = (6, 9)$, $m = 5$

10. $p = (^-1, 2)$, $m = ^-1$

11. $p = (^-2, 9)$, $m = 3$

12. $p = (^-2, ^-1)$, $m = 2$

13. $p = (3, 10)$, $m = ^-5$

14. $p = (7, 8)$, $m = 4$

Graphing Linear Inequalities

Graph the line $y > 2x + 3$.

1. $m = \frac{2}{1}$ $b = 3$
2. If $>$ or $<$, connect the points with a dotted line. 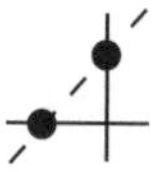
3. If $\geq$ or $\leq$, connect the points with a solid line. 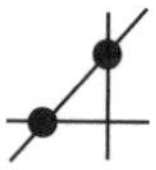
   The coordinate plane is now divided into 2 regions.
4. Test any (x, y) on each side of the line in the original inequality.

$y > 2x + 3$
Test point A $(^-1, 4)$.
Is $4 > 2(^-1) + 3$?
$4 > ^-2 + 3$
$4 > 1$ ⟶ true
(Shade this region)
Test point A.

$y > 2x + 3$
Test point B $(0, 0)$.
Is $0 > (0) + 3$?
$0 > 0 + 3$
$0 > 3$ ⟶ false
(Do not shade this region)
Test point B.

Test point A.

Test point B.

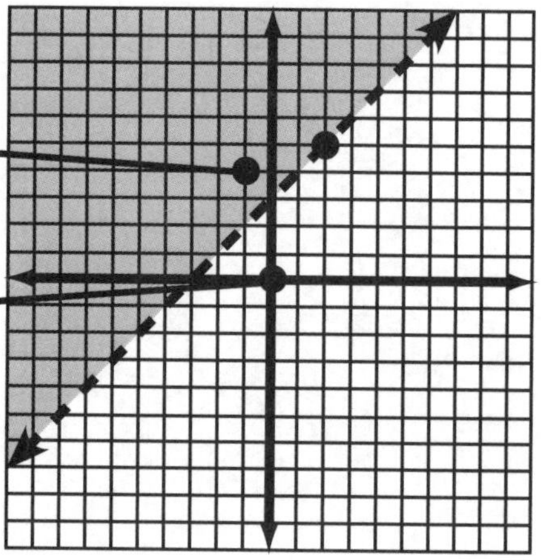

Graph the solution set.

1. $x + 4y > 8$

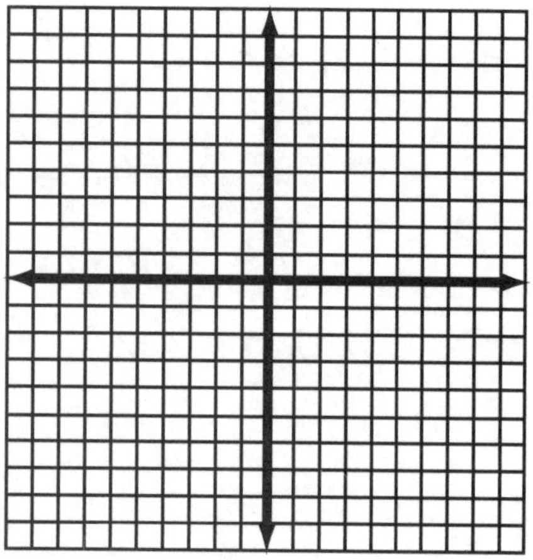

2. $^-2x + 2y \geq 10$

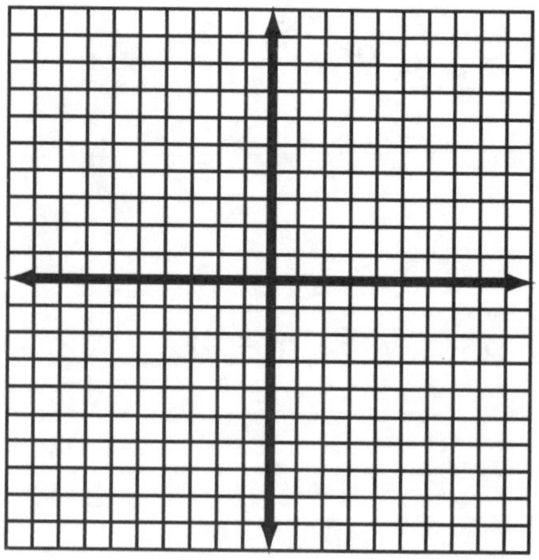

Name ______________________________ HSA-REI.D.12

Graphing Linear Inequalities

Graph the solution set.

1. $3x + 2y \geq 6$

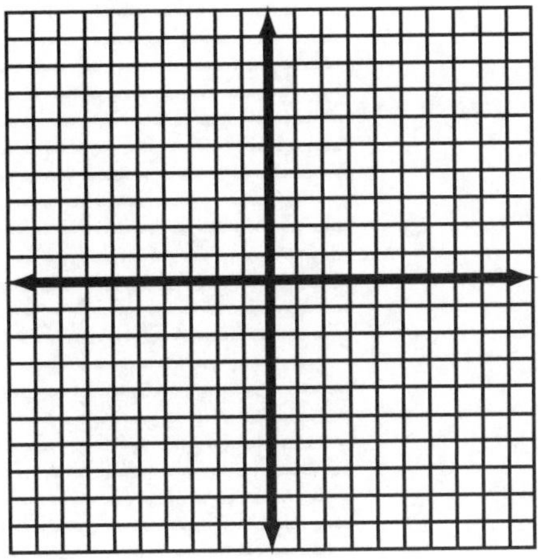

2. $2x + y < 4$

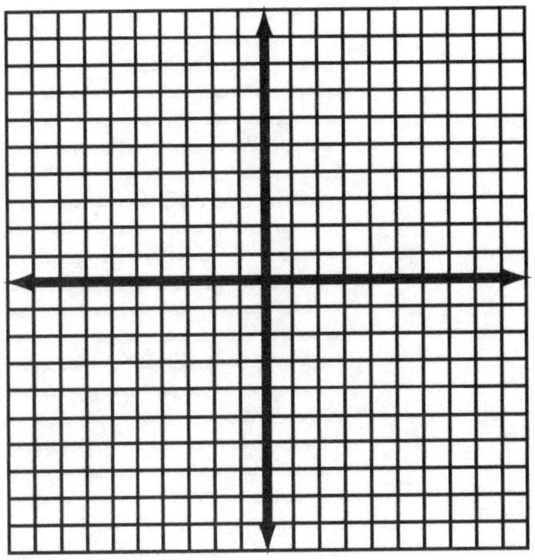

3. $6x - 3y > 15$

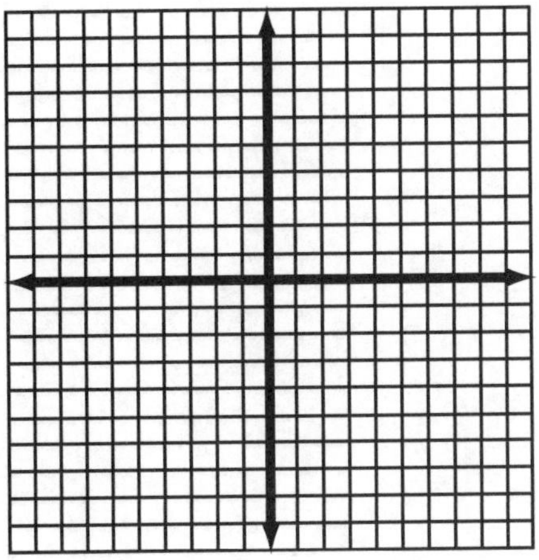

4. $2x + 3y \geq 6$

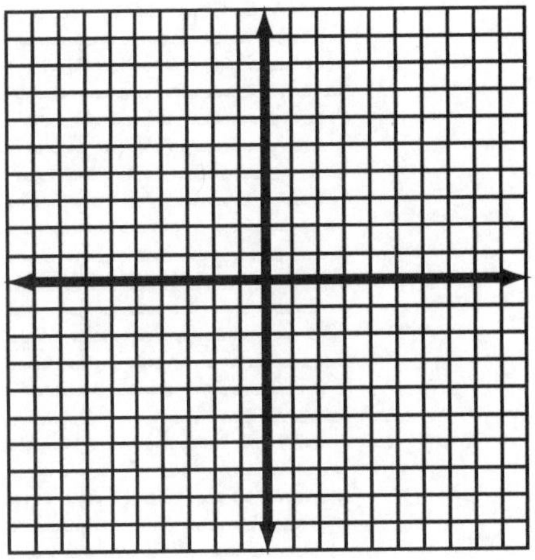

5. $3x - 4y \leq 12$

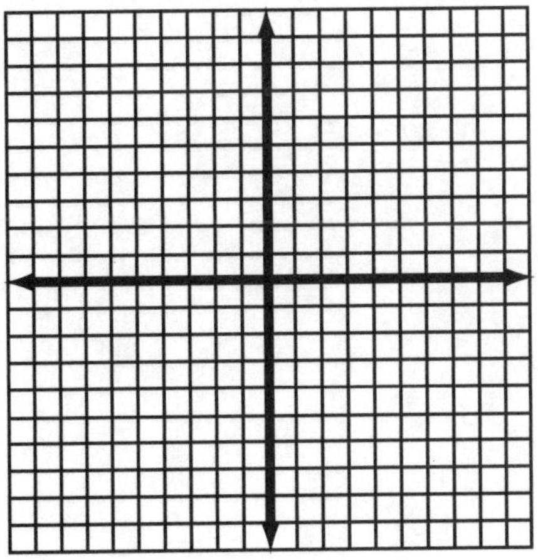

6. $4x + 2y < 6$

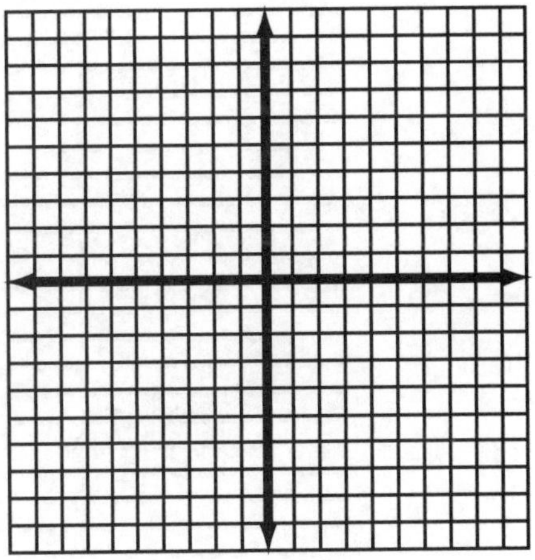

Name ____________________ HSA-REI.D.12

Graphing Linear Inequalities

Graph the solution set.

1. $x + 2y < 0$

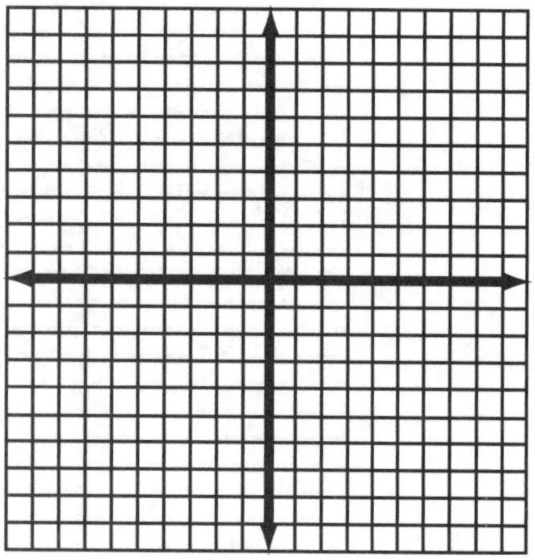

2. $5x - 2y \leq 10$

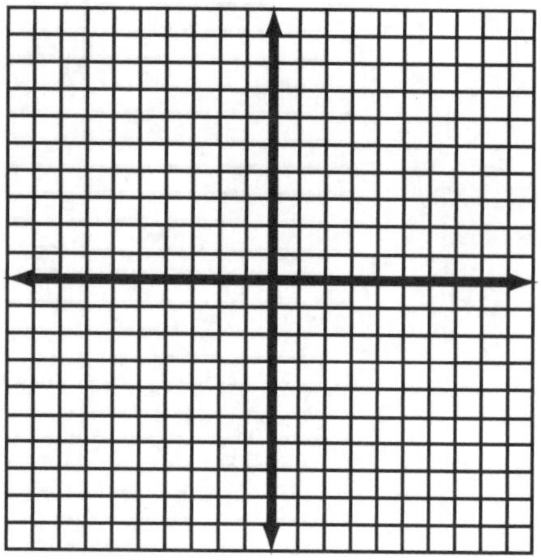

3. $6x - 3y > 18$

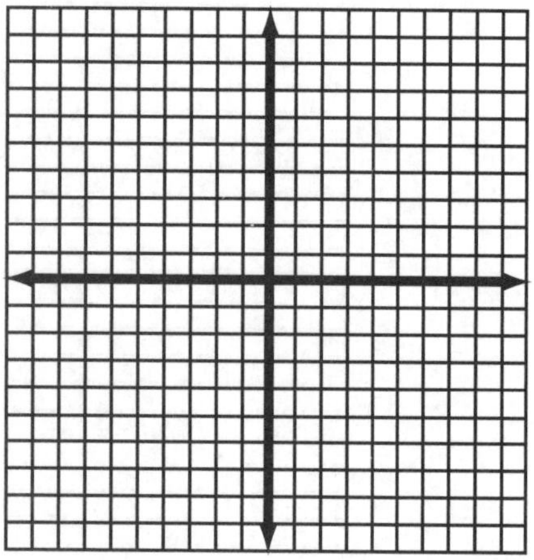

4. $2x - 5y < 10$

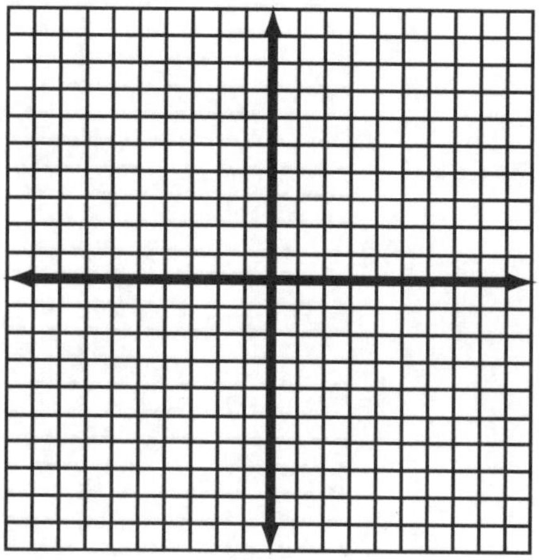

5. $^{-}5x + 5y \leq 15$

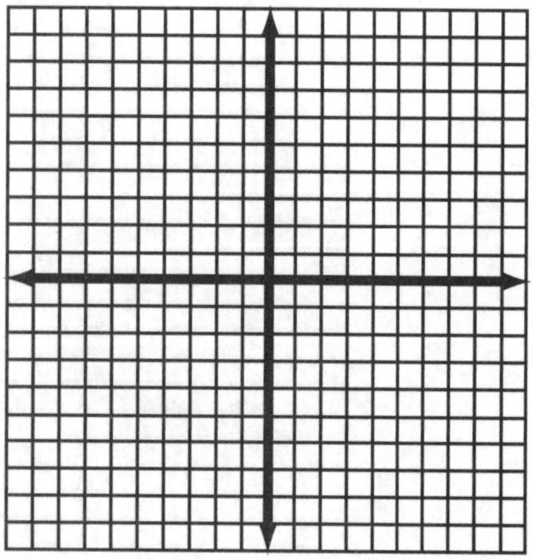

6. $y + 6 > 0$

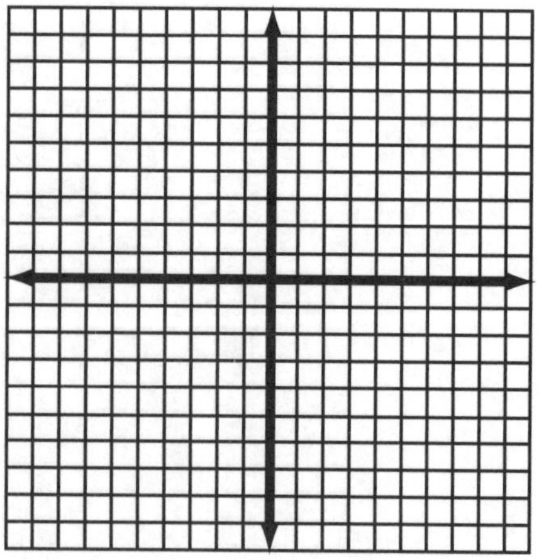

Name ______________________________ 8.EE.C.8b, HSA-REI.C.6, HSA-REI.D.11

Solving Systems of Equations by Graphing

When two or more equations are considered together, it is called a **system of equations**. The following example is a system of two linear equations in two variables.

$x + 2y = 4$

$2x + y = {}^{-}1$

The graphs of these equations are straight lines.

An ordered pair that is a solution of each equation of the system is a solution of the system of equations in two variables. The solution of a system of linear equations can be found by graphing the lines of the system. The solution of the system of equations is the point where the lines intersect.

Solve by graphing:

$x + 2y = 4$

$2x + y = {}^{-}1$

Graph each line and find the point of intersection.

The solution is $({}^{-}2, 3)$ because the ordered pair lies on each line.

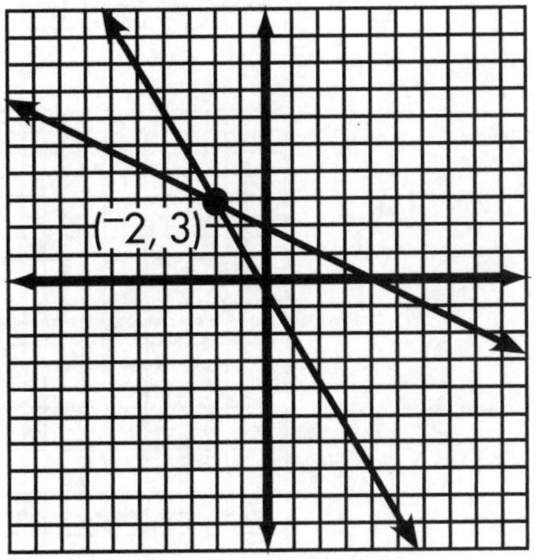

Solve by graphing.

1. $x + y = 7$

   $3x - y = {}^{-}3$

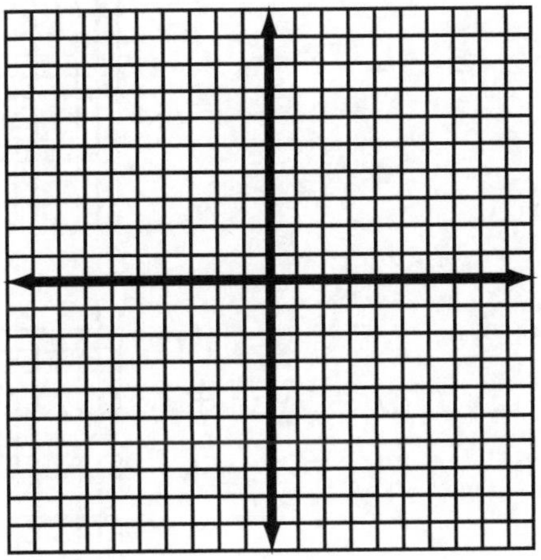

2. $x + y = 4$

   $x - y = 6$

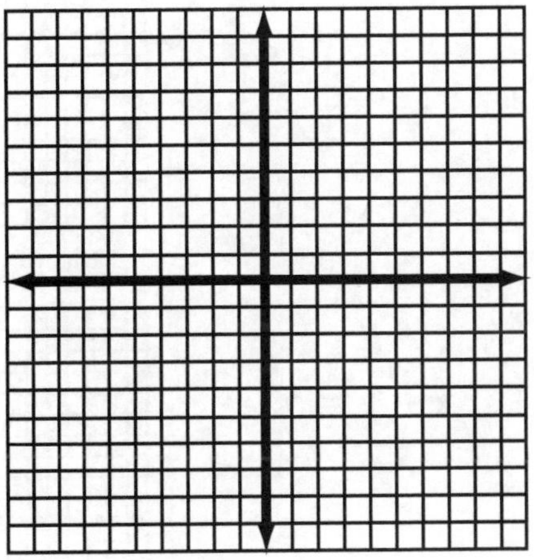

Name ______________________________ 8.EE.C.8b, HSA-REI.C.6, HSA-REI.D.11

Solving Systems of Equations by Graphing

Solve by graphing.

1. $3x - 4y = 12$
   $2x + 4y = {}^{-}12$

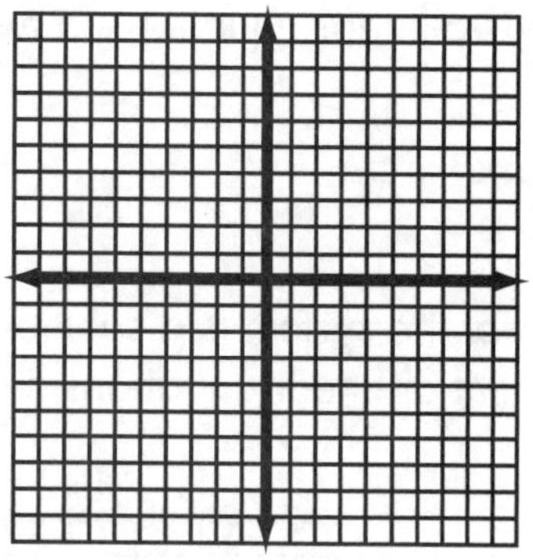

2. $4x - 2y = 8$
   $y = 2$

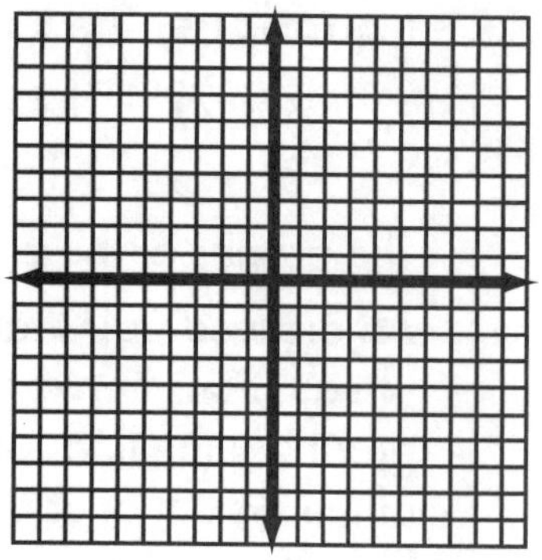

3. $x - y = 1$
   $x + 2y = 10$

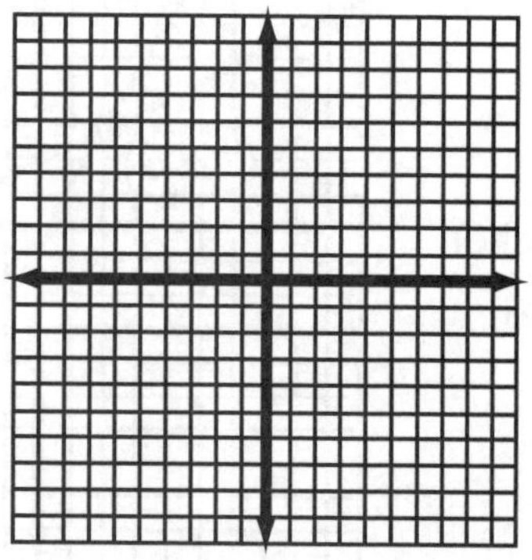

4. $x = 5$
   $y = {}^{-}1$

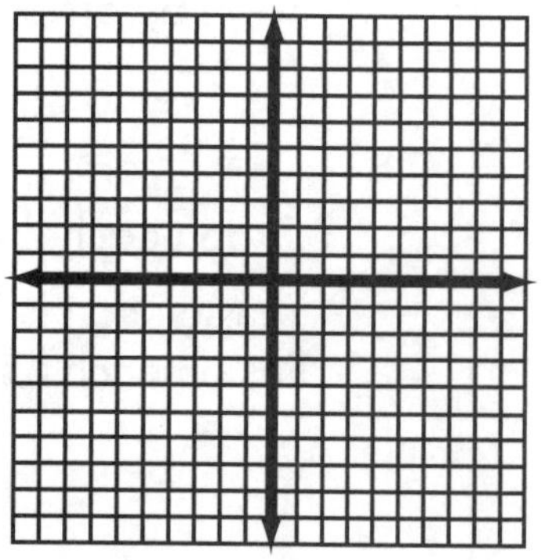

5. $y = 2x + 4$
   $y = x + 6$

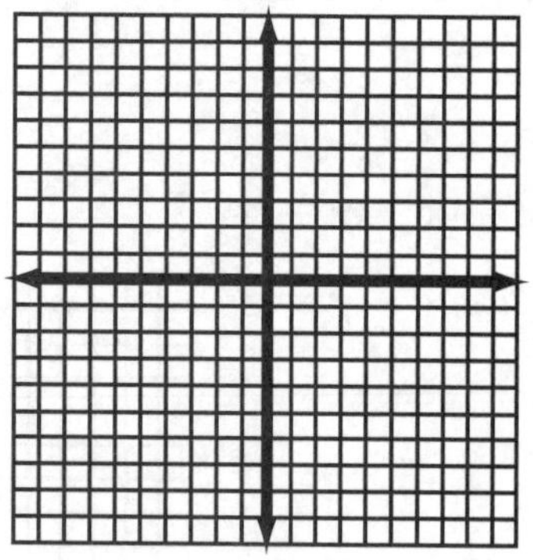

6. $x - y = {}^{-}4$
   $3x - y = {}^{-}12$

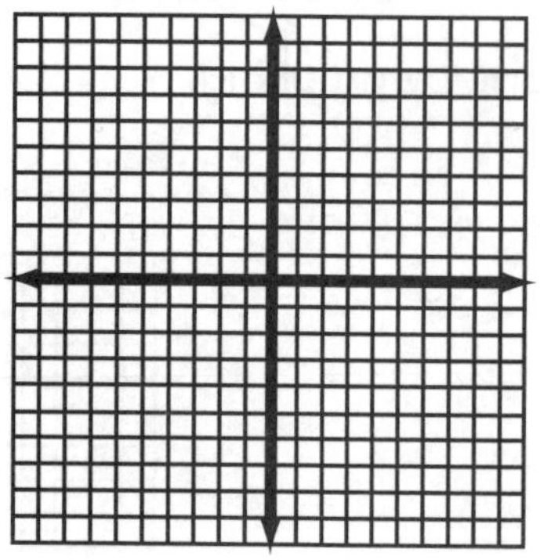

7. $x + y = 3$
   $x - y = 5$

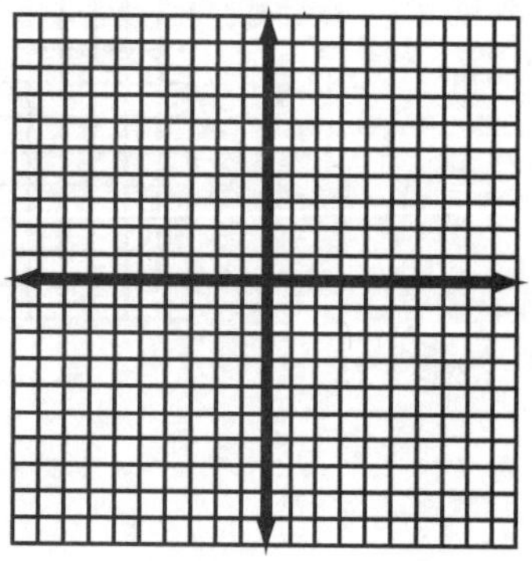

8. $x = 4$
   $6x - 2y = 4$

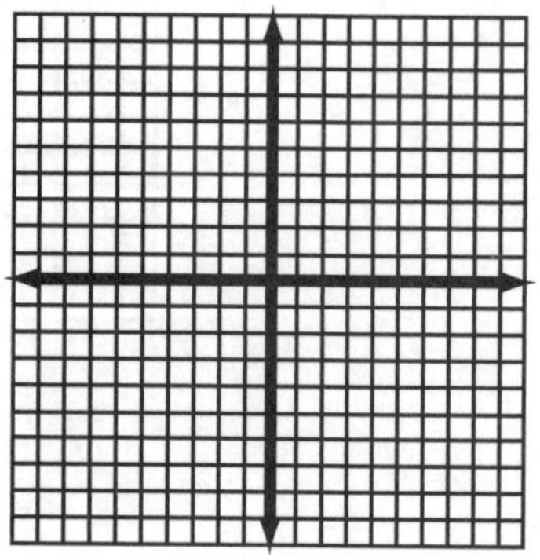

Name ______________________________ 8.EE.C.8b, HSA-REI.C.6, HSA-REI.D.11

Solving Systems of Equations by Graphing

Solve by graphing.

1. $y = x - 3$
   $x = 2y + 6$

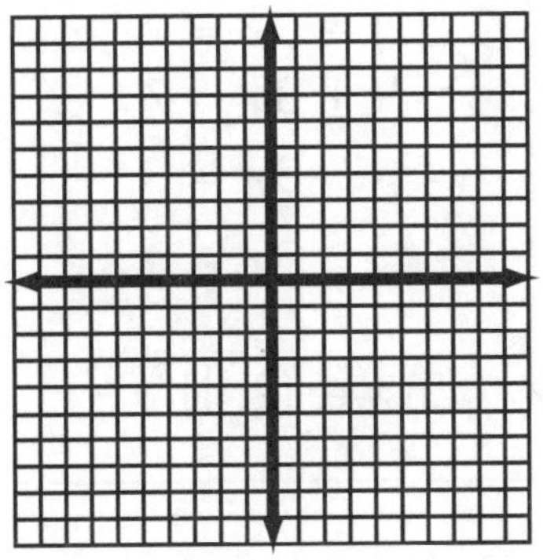

2. $-2x + y = 0$
   $6x + y = -16$

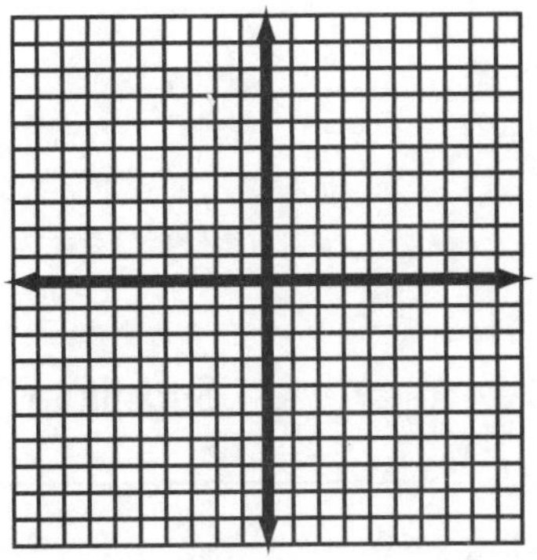

3. $y = 3x + 5$
   $y = 1 - x$

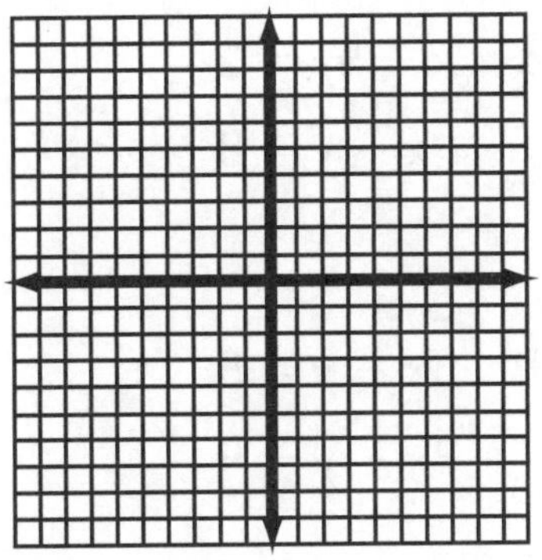

4. $y = x - 5$
   $y + 2x = 1$

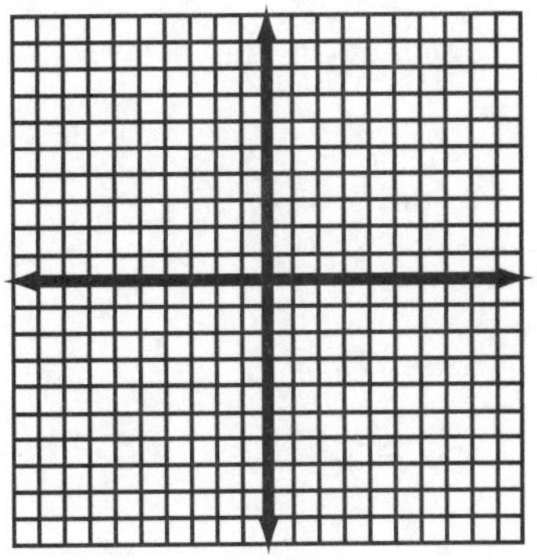

5. $x - 4y = 2$
   $x + 2y = 8$

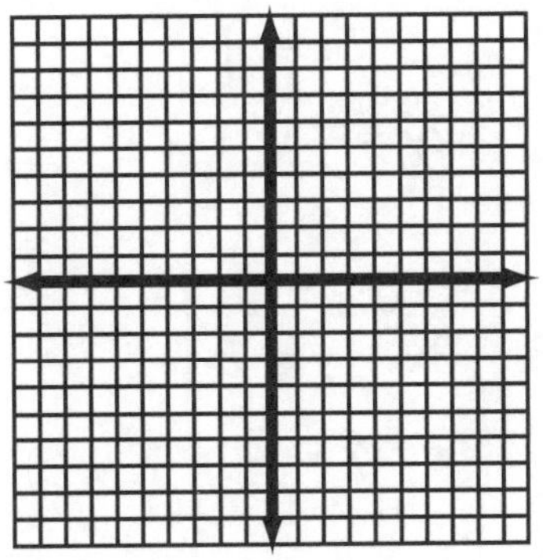

6. $x + y = 6$
   $3x - y = 2$

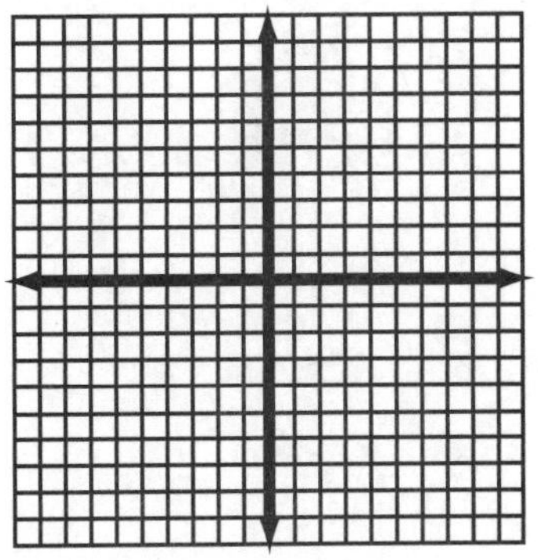

7. $2x + y = 6$
   $3x + y = 12$

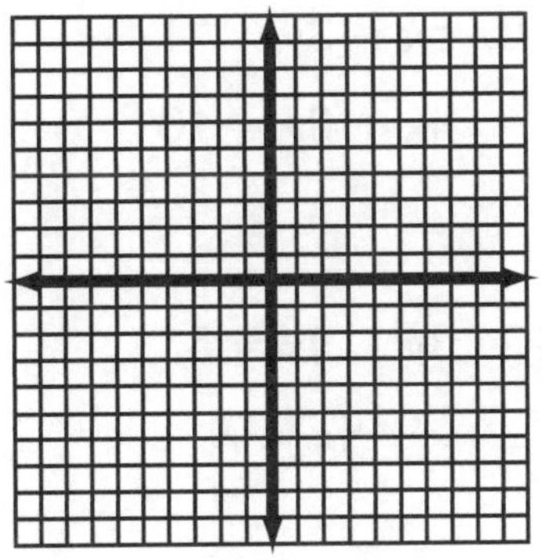

8. $y = x + 3$
   $2y = 3x + 1$

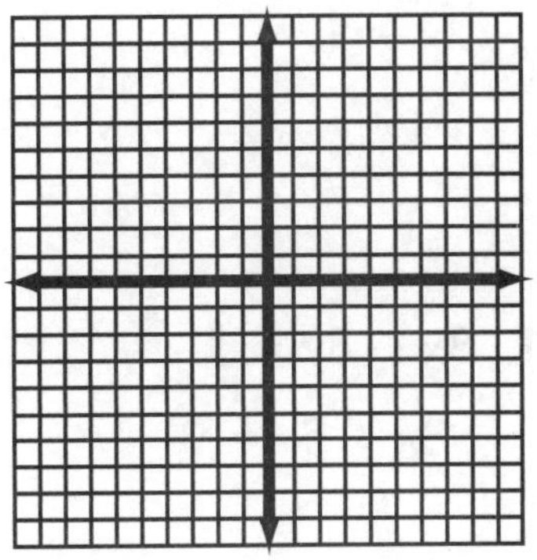

Name ____________________

8.EE.C.8b, HSA-REI.C.6

Solving Systems of Equations

$$\begin{array}{rl} 2x + 3y &= 7 \\ + \quad x - 3y &= {}^{-}1 \\ \hline 3x \qquad &= 6 \\ \frac{3x}{3} &= \frac{6}{3} \\ \mathbf{x} &= \mathbf{2} \end{array}$$

Substitute 2 for x in $2x + 3y = 7$.

$$\begin{array}{rl} 2x + 3y &= 7 \\ 2(2) + 3y &= 7 \\ 4 + 3y &= 7 \\ 3y &= 3 \\ \mathbf{y} &= \mathbf{1} \end{array}$$

answer (2, 1)

Solve.

1. $2x - 5 = y$
   $x - 7 = {}^{-}y$

2. $x + 4y = 2$
   ${}^{-}x + y = 8$

3. $y = 2x - 2$
   ${}^{-}y = x$

4. $3x + y = 8$
   $3x - y = 4$

5. $2x - y = 6$
   $3x + y = 4$

6. $y = 5x + 1$
   $2y = {}^{-}5x + 2$

7. $x + y = 7$
   $x - y = 3$

8. $3x + y = 5$
   $x - y = 7$

9. $3x - 4y = 14$
   $x + 4y = 2$

10. $5x - 3y = {}^{-}1$
    $4x + 3y = 10$

11. $8x - 3y = 1$
    ${}^{-}8x + 5y = 9$

12. $3y - 4x = 5$
    $y + 4x = 7$

Name ________________________________

8.EE.C.8b, HSA-REI.C.6

Solving Systems of Equations

1. Solve.

$6x + 5y = 6$

$6x - 3y = 6$

2. $\left.\begin{array}{r} 6x + 5y = 6 \\ (^-1)(6x - 3y = 6) \end{array}\right\}$ Multiply to create the additive inverse.

3. $\left.\begin{array}{r} 6x + 5y = 6 \\ ^-6x + 3y = {^-6} \end{array}\right\}$ Use the addition method.

$8y = 0$

$y = 0$

4. $\left.\begin{array}{r} 6x + 5y = 6 \\ 6x + 5(0) = 6 \end{array}\right\}$ Substitute.

$6x = 6$

$x = 1$

Solve.

1. $3x + 8y = 8$
   $2x + y = 1$

2. $3x - 4y = 4$
   $x - y = 1$

3. $4x - 4y = 16$
   $2x + 2y = 4$

4. $2x - 5y = 21$
   $x + y = 7$

5. $4x + 8y = 20$
   $x - y = 2$

6. $6x + 4y = {^-22}$
   $3x + 6y = {^-7}$

7. $2x + y = 3$
   $x + 3y = 4$

8. $3x + 3y = {^-3}$
   $x - y = {^-5}$

9. $4x + 2y = 10$
   $x + 2y = 1$

10. $3x + 5y = {^-8}$
    $x + 7y = {^-8}$

11. $x - y = 7$
    $x + 2y = 1$

12. $2x - 8y = 6$
    $^-x + 7y = 9$

Name ______________________

8.EE.C.8b, HSA-REI.C.6

Solving Systems of Equations

Solve.

1. $y = 3 - 2x$
   $y = 2 - 3x$

2. $x + y = 5$
   $x = y + 7$

3. $x - y = 1$
   $2x + y = 8$

4. $3x - y = 9$
   $y = x + 5$

5. $3x + 4y = 26$
   $^{-}2x + y = 1$

6. $y = 2x + 3$
   $y = 4x + 4$

7. $2x + 7y = 8$
   $x + 5y = 7$

8. $y = 4x + 4$
   $y = 2x + 8$

9. $x + 3y = 17$
   $2x + 3y = 22$

10. $4x - 7y = 9$
    $y = x - 3$

11. $8x - 5y = 9$
    $y = 2x - 4$

12. $2x + 4y = {}^{-}2$
    $3x + y = 7$

13. $3x + y = 5$
    $2x + 3y = 8$

14. $2x + 6y = 24$
    $x - 4y = {}^{-}2$

Name ______________________________ HSN-RN.A.2

Operations with Radicals

Simplify: $\sqrt{49x^4y^8} = 7\sqrt{x^4y^8} = 7x^2y^4$

Multiply: $\sqrt{3a} \bullet \sqrt{4a} = \sqrt{12a^2} = \sqrt{3 \bullet 4 \bullet a^2} = 2a\sqrt{3}$

Divide: $\sqrt{\frac{18}{2}} = \sqrt{9} = 3$ $\sqrt{\frac{9}{25}} = \frac{\sqrt{9}}{\sqrt{25}} = \frac{3}{5}$

Add or subtract: $2\sqrt{y} + 3\sqrt{y} + \sqrt{y} = 6\sqrt{y}$ $\sqrt{4x} + 3\sqrt{x} = 2\sqrt{x} + 3\sqrt{x} = 3\sqrt{x}$

Simplify.

1. $\sqrt{x^2y^{10}} =$

2. $\sqrt{27x^8} =$

3. $\sqrt{x^{16}} =$

4. $\sqrt{125b^{15}} =$

5. $\sqrt{x^2y^4} \bullet 2\sqrt{xy} =$

6. $\sqrt{9} \bullet \sqrt{32} =$

7. $3\sqrt{5} \bullet 2\sqrt{4} =$

8. $2\sqrt{4x^3y} \bullet y\sqrt{x^5y^7} =$

9. $\sqrt{\frac{36}{9}} =$

10. $\sqrt{\frac{27x}{3x}} =$

11. $3\sqrt{x^3} - 4\sqrt{x^3} =$

12. $3\sqrt{2y} + 2\sqrt{2y} =$

13. $2\sqrt{y} - 4\sqrt{y} =$

14. $3y\sqrt{2y} - y\sqrt{2y} =$

Name ______________________________ HSN-RN.A.2

Operations with Radicals

Simplify.

1. $\sqrt{x^{14}y^{6}} =$

2. $\sqrt{169y^{12}} =$

3. $\sqrt{16x^{4}} =$

4. $\sqrt{8x^{3}} =$

5. $\sqrt{81x^{6}} =$

6. $\sqrt{25x^{6}} =$

7. $4\sqrt{9x^{3}} \bullet 3\sqrt{4x} =$

8. $x\sqrt{5x^{3}y} \bullet x\sqrt{5x^{2}y} =$

9. $2\sqrt{9x^{2}} \bullet 2\sqrt{4x^{2}} =$

10. $6\sqrt{9xy} \bullet 4\sqrt{2xy} =$

11. $2\sqrt{4x^{3}y} \bullet 3\sqrt{3a^{2}b^{2}} =$

12. $\sqrt{\frac{x^{2}}{25}} =$

13. $\sqrt{\frac{8x^{3}}{2x}} =$

14. $\sqrt{\frac{18x^{3}}{2x}} =$

15. $x\sqrt{27} + x\sqrt{12} =$

16. $3\sqrt{6x} + 5\sqrt{6x} =$

17. $4\sqrt{2x^{3}} + 3\sqrt{2x^{3}} =$

18. $2\sqrt{50} - 4\sqrt{8} - 3\sqrt{72} =$

Name ______________________ HSN-RN.A.2

Operations with Radicals

Simplify.

1. $\sqrt{x^9y^9} =$

2. $\sqrt{54x^8} =$

3. $\sqrt{9a^4b^8} =$

4. $\sqrt{49x^4y^2} =$

5. $5\sqrt{2x^6y} \bullet 3\sqrt{3x^3y^5} =$

6. $4\sqrt{8a^6b} \bullet 4\sqrt{8a^4b^4} =$

7. $3\sqrt{2x^3} \bullet 3\sqrt{3x^2y^2} =$

8. $5\sqrt{4a} \bullet 2\sqrt{6a} =$

9. $x\sqrt{3x} \bullet \sqrt{3x^3} =$

10. $\sqrt{2x^4} \bullet \sqrt{10x^2y^2} =$

11. $\sqrt{\frac{9}{64}} =$

12. $\sqrt{\frac{50x^2}{2}} =$

13. $\sqrt{\frac{49x^2}{25x^3}} =$

14. $\sqrt{\frac{12x^2}{60}} =$

15. $\sqrt{\frac{3x^7}{108y^2}} =$

16. $4\sqrt{y^3} - 2\sqrt{y^3} =$

17. $y\sqrt{y^4} - 2y\sqrt{y^4} =$

18. $4x\sqrt{x^3} + 2x\sqrt{x^3} =$

19. $4\sqrt{x} - 2\sqrt{x} - 3\sqrt{x} + 5\sqrt{x} =$

20. $3\sqrt{4x^2y} - 8y\sqrt{y} =$

Answer Key

Name____________________ (7.NS.A.1d)

Adding Real Numbers

A number line is a great way to model the addition of real numbers.

−9−8−7−6−5−4−3−2−1 0 1 2 3 4 5 6 7 8 9

Add 3 + 5. (Start at 3 and move 5 places to the right since 5 is positive.) The answer is 8.
Add 3 + (−5). (Start at 3 and move 5 places to the left since 5 is negative.) The answer is −2.
When adding a number that is positive, move to the right. When adding a number that is negative, move to the left.

Explain how to add the following numbers using a number line, and tell the result.

1. 7 + 8 **Start at 7 and move 8 places to the right. 15**
2. −5 + 9 **Start at −5 and move 9 places to the right. 4**
3. 10 + (−7) **Start at 10 and move 7 places to the left. 3**
4. −11 + (−6) **Start at −11 and move 6 places to the left. −17**

Use the number line to add the numbers.

5. 4 + 9 **13**
6. −5 + 13 + (−11) **−3**
7. −7 + (−6) **−13**
8. 14 + 10 + (−3) **21**
9. 0 + (−5) **−5**
10. −6 + (−9) + 10 **−5**
11. −3 + 3 **0**
12. 8 + (−2) + (−4) + 6 **8**
13. 7 + (−12) **−5**
14. −11 + 5 + (−13) + 8 **−11**
15. −11 + 8 + (−1) **−4**
16. 12 + (−6) + (−14) + 17 **9**

Name____________________ (7.NS.A.1d)

Adding Real Numbers

−7 + 6 = −1

Add.

1. −55 + (−8) + (−4) + 54 = **−13**
2. 1 + (−5) + (−5) + 1 = **−8**
3. 2.7 + (−4.8) = **−2.1**
4. 3.54 + 4.27 + 7.43 = **15.24**
5. 10 + 7 + (−7) + (−10) = **0**
6. 16 + 21 + (−3) + 7 = **41**
7. 10 + 7 + (−16) + 9 + (−30) = **−20**
8. 5.8 + 8.4 = **14.2**
9. 2.76 + (−6.56) + (−9.72) = **−13.52**
10. 8 + (−7) = **1**
11. $2\frac{3}{5} + 4\ 3\frac{3}{7} =$ $\mathbf{7\frac{1}{35}}$
12. $-8\frac{3}{5} + 3\frac{3}{7} =$ $\mathbf{-5\frac{6}{35}}$
13. 2 + 5 + (−3) = **4**
14. $-5\frac{3}{4} + (-2\frac{3}{4}) + 8 =$ $\mathbf{-\frac{1}{2}}$
15. 7.3 + (3.9) = **11.2**
16. −21 + 12 + (−1) + (−17) = **−27**
17. 7.867 + (−5.329) = **2.538**
18. $-2\frac{3}{5} + (-5\frac{3}{7}) + 3 =$ $\mathbf{-5\frac{1}{35}}$
19. 3 + 12 + (−13) + 36 = **38**
20. $-3\frac{1}{6} + (-9\frac{3}{12}) + 6 =$ $\mathbf{-6\frac{5}{12}}$

Name____________________ (7.NS.A.1d)

Adding Real Numbers

Add.

1. $2\frac{3}{5} + (-3\frac{2}{5}) + {}^-6 =$ $\mathbf{-6\frac{4}{5}}$
2. 21 + 9 + (−6) + 7 = **31**
3. 12 + (−9) + 17 = **20**
4. 2.54 + −5.87 + −32.65 = **−35.98**
5. 1.45 + 2.65 + (−9.43) = **−5.33**
6. 21 + 3 + (−13) + 22 = **33**
7. 3 + (−3) + 4 + (−5) = **−1**
8. 3.3 + (−3.4) + 5.5 = **5.4**
9. 3.6 + (−2.5) + −5.5 = **−4.4**
10. −0.6 + (−0.56) + 3 = **1.84**
11. $3\frac{5}{8} + (-1\frac{3}{4}) + 2 =$ $\mathbf{3\frac{7}{8}}$
12. 4.524 + 7.342 = **11.866**
13. $-7\frac{2}{4} + 2\frac{3}{4} =$ **−4 3/4**
14. 34 + (−13) + 18 + 0 + 34 = **73**
15. 8.43 + (−10.98) + (−3.23) = **−5.78**
16. 2.54 + (−5.21) + (−6.34) = **−9.01**
17. $-2\frac{1}{3} + (-5\frac{7}{10}) + (-7) =$ $\mathbf{-15\frac{1}{30}}$
18. $-1\frac{2}{3} + (-3\frac{3}{5}) + 4 =$ $\mathbf{-1\frac{4}{15}}$
19. $2\frac{1}{2} + 6\frac{1}{2} =$ **9**
20. $4\frac{3}{5} + (-3\frac{2}{5}) + (-8) =$ $\mathbf{-6\frac{4}{5}}$

Name____________________ (7.NS.A.1c, 7.NS.A.1d)

Subtracting Real Numbers

For all real numbers *a* and *b*, $a - b = a + (^-b)$. Simply stated: To subtract a number, add its opposite.

7 − 10 = 7 + (−10) = −3	5 − 8 + 3 − 1 = 5 + (−8) + 3 + (−1) = 5 + 3 + (−8) + (−1) = 8 + (−9) = −1 (group the numbers with like signs)	−4 − (−12) = −4 + 12 = 8

Change each problem into an addition problem.

1. 7 − 9 **7 + (−9)**
2. −6 − (−4) **−6 + 4**
3. −11 − 5 **−11 + (−5)**
4. 12 − (−15) **12 + 15**
5. 8 − 3 **8 + (−3)**
6. 22 − 5 **22 + (−5)**
7. 4 − 11 **4 + (−11)**
8. −4 − (−9) **−4 + 9**

Subtract.

9. 9 − 11 **−2**
10. 0 − (−12) **12**
11. −5 − 4 **−9**
12. 6 − (−6) **12**
13. −1 − (−1) **0**
14. −7 − 6 **−13**
15. 3 − (−5) **8**
16. 17 − 23 **−6**

Answer Key

Name ______________________ (7.NS.A.1c, 7.NS.A.1d)

Subtracting Real Numbers

$10 - (^-4) = 10 + 4 = 14$

Subtract.

1. $9 - (^-32) =$ **41**
2. $^-99 - (^-42) =$ **-57**
3. $4 - (^-8) =$ **12**
4. $0 - 21 =$ **-21**
5. $45 - 301 =$ **-256**
6. $^-19 - 8 =$ **-27**
7. $^-43 - 6 =$ **-49**
8. $9 - (^-2) - 8 - 7 =$ **-4**
9. $35 - 67 - 85 - 21 - 12 =$ **-150**
10. $12 - 7 - (^-16) - 9 - (^-34) =$ **46**
11. $18 - (^-13) =$ **31**
12. $121 - 45 =$ **76**
13. $-\frac{4}{7} - \frac{1}{3} - (\frac{2}{3}) =$ $\mathbf{-1\frac{4}{7}}$
14. $3.434 - 7.294 =$ **-3.86**
15. $8 - 2.8 =$ **5.2**
16. $8 - (^-14) =$ **22**
17. $3.9 - 4.9 =$ **-1**
18. $^-7 - (^-3) =$ **-4**
19. $2.19 - 7.8 - 8.31 =$ **-13.92**
20. $38 - 39 - (^-13) =$ **12**

Name ______________________ (7.NS.A.1c, 7.NS.A.1d)

Subtracting Real Numbers

Subtract.

1. $^-9 - (^-5) =$ **-4**
2. $321 - (^-34) =$ **355**
3. $\frac{2}{3} - \frac{4}{5} =$ $\mathbf{-\frac{2}{15}}$
4. $\frac{3}{5} - \frac{7}{8} =$ $\mathbf{-\frac{11}{40}}$
5. $5.34 - 9.9 - 3.65 =$ **-8.21**
6. $9.432 + 4.348 - 44.938 =$ **-31.158**
7. $245 - 32 - (^-36) =$ **249**
8. $44 - 35 - 34 - 32 =$ **-57**
9. $8 - (^-5) - 7 - 9 =$ **-3**
10. $43 - 88 - 35 - 21 =$ **-101**
11. $-\frac{2}{5} - \frac{3}{4} - (-\frac{4}{5}) =$ $\mathbf{-\frac{7}{20}}$
12. $^-45 - 5 =$ **-50**
13. $-\frac{2}{3} - \frac{1}{3} - (-\frac{1}{3}) =$ $\mathbf{-\frac{2}{3}}$
14. $-\frac{4}{5} - \frac{1}{2} - \frac{2}{5} =$ $\mathbf{-1\frac{7}{10}}$
15. $4 - 12.9 =$ **-8.9**
16. $7 - (^-33) =$ **40**
17. $3.4 - 7.4 =$ **-4**
18. $2.456 - 4.345 - 5.457 =$ **-7.346**
19. $23 - (^-21) =$ **44**
20. $4.346 - 0.4537 =$ **3.8923**

Name ______________________ (7.NS.A.2a, 7.NS.A.2c)

Multiplying Real Numbers

The **property of zero for multiplication** states that for all real numbers *a*, $a \bullet 0 = 0$ and $0 \bullet a = 0$. Simply stated, any real number multiplied by 0 is 0. For example, $0 \bullet 20 = 0$.

To multiply two real numbers with the same signs:
1. Multiply their absolute values.
2. The sign of their product is positive.

positive • positive = positive (+) (+) (+) $3 \bullet 12 = 36$

negative • negative = positive (−) (−) (+) $^-7 \bullet {}^-8 = 56$

To multiply two real numbers with different signs:
1. Multiply their absolute values.
2. The sign of their product is negative.

negative • positive = negative (−) (+) (−) $^-2 \bullet 5 = {}^-10$

positive • negative = negative (+) (−) (−) $4 \bullet {}^-8 = {}^-32$

Write the sign of the product for each number.

1. $(^-10)4$ **−**
2. $8(^-1)$ **−**
3. $(^-2)(^-3)$ **+**
4. $(7)(5)(^-3)$ **−**
5. $5(6)$ **+**
6. $(^-2)(^-7)$ **+**
7. $(^-12)(^-4)(^-1)$ **−**
8. $(^-6)(4)(^-2)$ **+**

Multiply to find each product.

9. $4(6)(^-1)$ **-24**
10. $(^-1)(^-4)(^-3)$ **-12**
11. $(-\frac{1}{2})(2)$ **-1**
12. $(7)(^-3)(0)$ **0**
13. $(5)(3)$ **15**
14. $(-\frac{1}{8})(^-16)(4)$ **8**
15. $(^-9)(^-4)$ **36**
16. $(^-7)(7)$ **-49**

Name ______________________ (7.NS.A.2a, 7.NS.A.2c)

Multiplying Real Numbers

$(^-2)(^-3) = 6$

Multiply.

1. $4 \bullet 9 =$ **36**
2. $^-4 \bullet 12 =$ **-48**
3. $(-\frac{5}{9})(8.8) =$ $\mathbf{4.\overline{8}}$
4. $(^-3)(0) =$ **0**
5. $(^-3)(^-9) =$ **27**
6. $6(23) =$ **138**
7. $(12)(^-3)(4) =$ **-144**
8. $(^-5)(^-5)(^-5) =$ **-125**
9. $(5)(2)(^-1) =$ **-10**
10. $(7)(^-9)(^-12) =$ **756**
11. $(-\frac{2}{3})(^-1.6) =$ $\mathbf{1.0\overline{6}}$
12. $^-7(^-7) =$ **49**
13. $(54.2)(^-3.55) =$ **-192.41**
14. $(2.22)(^-1.11) =$ **-2.4642**
15. $(7.44)(3.2)(4.3) =$ **102.3744**
16. $(2.4)(^-1.4) =$ **-3.36**
17. $(-\frac{3}{5})(\frac{3}{5}) =$ $\mathbf{-\frac{9}{25}}$
18. $(-\frac{4}{5})(2.2) =$ **-1.76**
19. $^-8 \bullet 12 =$ **-96**
20. $(0)(2)(^-213) =$ **0**

Answer Key

Name________________________ (7.NS.A.2a, 7.NS.A.2c)

Multiplying Real Numbers

Substitute and multiply.

$x = {}^-3, y = {}^-5, z = 0$

1. xy **15**
2. ^-3yz **0**
3. xyz **0**
4. $4x(^-2y)$ **-120**
5. $^-3(10xy)$ **-450**
6. $2y(5y)$ **250**

$x = {}^-2.4, y = 3.1, z = {}^-4.8$

7. $4xz$ **46.08**
8. $yz(^-z)$ **-71.424**
9. $^-5(xyz)$ **-178.56**
10. $^-x(^-6z)$ **69.12**
11. $(xy)(^-3x)$ **-53.568**
12. $(3yz)(^-7x)$ **-749.952**

$x = \frac{1}{2}, y = -\frac{2}{3}, z = \frac{3}{5}$

13. $8(xy)$ **$-2\frac{2}{3}$**
14. xyz **$-\frac{1}{5}$**
15. $^-12x(^-5yz)$ **-12**
16. $(xy)(3xz)$ **$-\frac{3}{10}$**
17. $(^-2z)(3y)(6x)$ **$7\frac{1}{5}$**
18. $^-z(10xyz)$ **$1\frac{1}{5}$**

 13

Name________________________ (7.NS.A.2b, 7.NS.A.2c)

Dividing Real Numbers

The **multiplicative inverse property** states that for each nonzero a, there is exactly one number $\frac{1}{a}$ such that: $a \bullet \frac{1}{a} = 1$ and $\frac{1}{a} \bullet a = 1$.

The number $\frac{1}{a}$ is called the reciprocal of a.

For example, the reciprocal of 9 is $\frac{1}{9}$ and the reciprocal of $\frac{1}{2}$ is 2. Simply flip the number to find its reciprocal.

To divide the number a by the number b, multiply a by the reciprocal of b.

$a \div b = a \bullet \frac{1}{b}$ (The result will be the quotient of a and b.)

$12 \div 4 = 12 \bullet \frac{1}{4} = 3$ $\quad 4x \div \frac{1}{4} = 4x \bullet 4 = 16x$ $\quad 15 \div \frac{3}{2} = 15 \bullet \frac{2}{3} = 10$

To divide two nonzero real numbers:

1. The quotient is positive if both numbers have the same sign.
2. The quotient is negative if both numbers have different signs.

$^-15 \div 5 = {}^-15 \bullet \frac{1}{5} = {}^-3$ $\quad 3 \div \frac{1}{4} = 3 \bullet 4 = 12$ $\quad 1 \div -\frac{4}{5} = 1 \bullet -\frac{5}{4} = -\frac{5}{4}$

Write the reciprocal of each number. Write none if it does not exist.

1. 2 **$\frac{1}{2}$**
2. 1 **1**
3. 0 **none**
4. $-\frac{1}{9}$ **-9**
5. $-\frac{1}{4}$ **-4**
6. $\frac{4}{3}$ **$\frac{3}{4}$**
7. $^-8$ **$-\frac{1}{8}$**
8. 10 **$\frac{1}{10}$**

Divide.

9. $\frac{9}{3}$ **3**
10. $\frac{^-28}{7}$ **-4**
11. $^-15 \div 3$ **-5**
12. $\frac{^-7}{8} \div \frac{1}{8}$ **-7**
13. $36 \div (^-4)$ **-9**
14. $\frac{^-8}{^-8}$ **1**
15. $44 \div (^-11)$ **-4**
16. $0 \div \frac{4}{5}$ **0**

14

Name________________________ (7.NS.A.2b, 7.NS.A.2c)

Dividing Real Numbers

$9 \div 4.5 = 2$

Divide.

1. $\frac{49}{7} =$ **7**
2. $90 \div 15 =$ **6**
3. $(^-12) \div (9.9) =$ **$-1.\overline{21}$**
4. $(-\frac{2}{3}) \div (^-18) =$ **$-\frac{1}{27}$**
5. $^-42 \div 7 =$ **-6**
6. $45 \div (^-8) =$ **-5.625**
7. $\frac{^-36}{4} =$ **-9**
8. $(-\frac{3}{5}) \div (\frac{3}{5}) =$ **-1**
9. $^-72 \div (9) =$ **-8**
10. $^-21 \div (^-9) =$ **$2.\overline{3}$**
11. $\frac{102}{17} =$ **6**
12. $0 \div (^-8) =$ **0**
13. $\frac{95}{5} =$ **19**
14. $\frac{63}{^-9} =$ **-7**
15. $(^-3.4) \div (^-9.99) =$ **$0.\overline{340}$**
16. $^-50 \div [40 \div (^-20)] =$ **25**
17. $(-\frac{4}{6}) \div (36) =$ **$-\frac{1}{54}$**
18. $56 \div (^-28 \div 7) =$ **-14**
19. $(^-38 \div 19) \div (^-2) =$ **1**
20. $^-45 \div [^-20 \div (^-4)] =$ **-9**

 15

Name________________________ (7.NS.A.2b, 7.NS.A.2c)

Dividing Real Numbers

Substitute and divide.

$x = {}^-24, y = 6, z = {}^-8$

1. $\frac{x}{y}$ **-4**
2. $\frac{yz}{^-x}$ **-2**
3. $\frac{^-z}{xy}$ **-18**
4. $\frac{4z}{5y}$ **$-1\frac{1}{15}$**
5. $\frac{^-xy}{^-yz}$ **3**
6. $\frac{2y}{^-5x}$ **$\frac{1}{10}$**

$w = 2.8, x = 1.4, y = {}^-8.4, z = {}^-11.2$

7. $(w) \div (^-x)$ **2**
8. $(z) \div (^-y)$ **-1.3**
9. $\frac{x}{(^-w)}$ **-0.5**
10. $(^-z) \div (^-w)$ **-4**
11. $\frac{^-3y}{2x}$ **9**
12. $(z) \div (^-x)$ **-8**

$x = -\frac{1}{4}, y = -\frac{1}{3}, z = \frac{4}{5}$

13. $(z) \div (x)$ **$-3\frac{1}{5}$**
14. $\frac{x}{^-z}$ **$\frac{5}{16}$**
15. $\frac{(^-2x)}{(y)}$ **$-1\frac{1}{2}$**
16. $(y) \div (^-xz)$ **$-1\frac{2}{3}$**
17. $(^-z) \div (y) \div (^-x)$ **$9\frac{3}{5}$**
18. $\frac{^-z}{xy}$ **$-9\frac{3}{5}$**

16

Answer Key

Name ____________________ 5.OA.A.1, 6.EE.A.2c

Order of Operations

When solving an equation, be sure to follow the **order of operations**.

1. Parentheses
2. Exponents
3. Multiplication & Division
4. Addition & Subtraction

$14 - 54 \div 6 = 14 - 9 = 5$

Solve.

1. $3 \times 15 \div 5 =$ **9**
2. $35 \div 5 - 9 =$ **−2**
3. $3 + 2 \times 4 =$ **11**
4. $5 \times 2 \times 8 =$ **80**
5. $6 - 40 \div 8 =$ **1**
6. $12 - 30 \div 6 =$ **7**
7. $32 \div 4 \times 3 =$ **24**
8. $8 + 3 \times 2 =$ **14**
9. $4 + 12 \div 2 =$ **10**
10. $9 + 20 \div 5 =$ **13**
11. $15 - 75 \div 5 =$ **0**
12. $9 - 3 + 6 =$ **12**
13. $2 \times 8 \div 4 =$ **4**
14. $3 + 3 - 2 =$ **4**

Name ____________________ 5.OA.A.1, 6.EE.A.1, 6.EE.A.2c

Order of Operations

Remember to follow the **order of operations**.

1. Parentheses
2. Exponents
3. Multiplication & Division
4. Addition & Subtraction

$(3^3 + 6 \times 5) - 2 = (27 + 6 \times 5) - 2 = (27 + 30) - 2 = 55$

Solve.

1. $(3^2 + 2 \times 3) \div 5 =$ **3**
2. $5^2 - 4^2 + 2 =$ **11**
3. $(4 + 2)^2 =$ **36**
4. $(11 - 8)^3 =$ **27**
5. $2(7 + 2) =$ **18**
6. $(9 - 7)^3 - (4 + 3) =$ **1**
7. $(14 - 6)2 =$ **16**
8. $4 + 3(12 - 9) =$ **13**
9. $5^2 - 2^3 =$ **17**
10. $3 \times 8 - (3 \times 2 + 7) =$ **11**
11. $(5^2 - 3 \times 5) \div 2 =$ **5**
12. $7 + 2^2(5 + 2) =$ **35**
13. $3 + 7^2 =$ **52**
14. $(2^2 + 3)^2 - 4 =$ **45**
15. $6 + 7 \times 3 - 9 \times 2 =$ **9**
16. $(2 \times 3) + (21 \div 7) =$ **9**
17. $7^2 - 2(3 \times 3 + 5) =$ **21**
18. $3 + (6 \times 2) =$ **15**

Name ____________________ 5.OA.A.1, 6.EE.A.1, 6.EE.A.2c

Order of Operations

Solve.

1. $8 - 4 \bullet 5(3 - 2) + 3 =$ **−9**
2. $12 \div (2 - 7) + 7 =$ **4.6**
3. $(14 - 9) + 4 =$ **9**
4. $\frac{3^2 - 5 \bullet 7 - 4^2}{(-4 - 7 - 12) + 8} =$ **$2\frac{4}{5}$**
5. $9(3 \div 3) + 4(^-5 \bullet 9) \div 3 =$ **−51**
6. $3 - (6 \bullet 6) - 3 \bullet 0 =$ **−33**
7. $36 \div 9 - 8 + 21 \div 3 =$ **3**
8. $5(3 - 8) \bullet 3 + 8 - 3 =$ **−70**
9. $3 \bullet 5 + 9 \bullet 7 =$ **78**
10. $\frac{(5 - 9)^2 + 2}{(7 - 8)^2 \bullet 3^2} =$ **2**
11. $4^2 + 3^2 - 7^2 =$ **−24**
12. $\frac{3^2 - 10}{4^2 - 12} =$ **$\frac{1}{4}$**
13. $8^2 - \frac{26}{(4 + 9)} + 4 =$ **66**
14. $\frac{5 \bullet 7 - (3 + 4)}{-(2^2) - 2^2 + 3^2} =$ **28**
15. $\frac{4 + 2 \bullet 3 + 4 - 3}{2^2 \bullet 3^2 - 3} =$ **$\frac{1}{3}$**
16. $\frac{3 + 10 - 19 + 32}{3^2 - 1 + 2^2} =$ **$2\frac{1}{6}$**
17. $12 \div [3 + (6 + 3)] =$ **1**
18. $3 \bullet (0 - 7) + 8 \div 2^2 =$ **−19**

Name ____________________ 6.NS.C.5, 6.NS.C.6a, 6.NS.C.7

Real-Number Operations with Absolute Value

Every real number has an opposite. Opposite numbers are the same distance from 0 on a number line and lie on the opposite sides of 0. The opposite of a positive number is a negative number. The opposite of a negative number is a positive number. The numbers 3 and −3 are opposites. Find their places on the real number line to the right. The additive inverse of a number is the same as the opposite of a number.

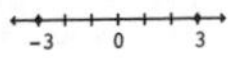

Remember, the opposite of 0 is simply 0 since it is neither positive nor negative.

The symbol $|x|$ is called the absolute value of x. The absolute value of a number is the distance between the number and 0 on a number line. The absolute value of a number, whether positive or negative, is always positive.

$|10| = 10$ $|^-2| = 2$ $|^-102| = 102$ $^-|12| = {}^-12$

Note: The answer to $^-|12|$ is $^-12$ because the absolute value of 12 is 12. But, it is multiplied by a negative, which resulted in $^-12$.

Write the opposite of each number.

1. 5 **−5**
2. −2.6 **2.6**
3. −40 **40**
4. 2.8 **−2.8**

Write a real number to represent each situation.

5. a gain of 12 yards **12**
6. a temperature drop of 8° **−8**
7. a deposit of $89.26 **89.26**
8. a withdrawal of $75 **−75**

Write the absolute value of each number.

9. $|10|$ **10**
10. $|23|$ **23**
11. $|0|$ **0**
12. $|^-42|$ **42**

Simplify.

13. $^-|^-15|$ **−15**
14. $|^-5| \bullet |^-4|$ **20**
15. $^-|4 \bullet 3|$ **−12**
16. $^-|30 \div 6|$ **−5**
17. $|^-8| + |8|$ **16**
18. $|7| - |^-7|$ **0**

Answer Key

Name_______________ (6.NS.C.6a, 6.NS.C.7)

Real-Number Operations with Absolute Value

The **absolute value** of a number is its distance from zero.
$^-|5 - 11| = ^-|^-6| = ^-6$ $|^-4| \cdot |^-3| = 4 \cdot 3 = 12$

Simplify.

1. $|^-3| =$ **3**
2. $|^-14| =$ **14**
3. $9 + |^-4| =$ **13**
4. $^-5 \cdot |4| + |5| =$ **−15**
5. $^-|4 \cdot 7| =$ **−28**
6. $|21| + 9 =$ **30**
7. $23 + |8| =$ **31**
8. $|12| - |^-15| =$ **−3**
9. $7 - |^-23| + |^-7| =$ **−9**
10. $|^-6| \cdot |8| =$ **48**
11. $^-|^-3 + 7| =$ **−4**
12. $^-|^-5 + 10| \div |^-3| =$ $-\frac{5}{3}$
13. $|^-9| + |23| =$ **32**
14. $|^-17| \div |^-17| =$ **1**
15. $|1| - |0| + 6 =$ **7**
16. $|^-67| - |^-17| =$ **50**
17. $|4| - |^-12 \div 4| =$ **1**
18. $|3 - 13| \div |^-5| =$ **2**
19. $|24| \div |^-12| =$ **2**
20. $^-|9| \cdot |^-9| =$ **−81**

Name_______________ (6.NS.C.6a, 6.NS.C.7)

Real-Number Operations with Absolute Value

Simplify.

1. $|^-4 + 8| =$ **4**
2. $^-|5| \cdot |^-7| =$ **−35**
3. $12 \div |^-3 + 7| =$ **3**
4. $^-|11| - |^-20| =$ **−31**
5. $|6 - 18| + |^-17 + 9| =$ **20**
6. $|^-27| \div |^-3| \cdot ^-|^-2| =$ **−18**
7. $^-48 \div |^-2 \cdot 6| =$ **−4**
8. $|^-83| - |^-38| =$ **45**
9. $|24| \cdot |^-4 + 1| =$ **72**
10. $10 - |^-13| + |^-7| =$ **4**
11. $|72 \div 8| \cdot |^-11| =$ **99**
12. $^-|16 - 21| - |^-7 + 4| =$ **−8**
13. $23 + |^-45 \div 9| - |52| =$ **−24**
14. $^-|^-4 - 15| =$ **−19**
15. $|^-83| + |^-16| =$ **99**
16. $^-|4| \cdot |^-4| =$ **−16**
17. $|^-7 + 15| \div |^-20 - 12| =$ $\frac{1}{4}$
18. $^-41 + |^-29| \cdot |0| =$ **0**
19. $|^-15| \div |3| - |45| =$ **−40**
20. $|22 \div 11| \cdot |^-32 \div 8| =$ **8**

Name_______________ (6.EE.A.2b, 6.EE.A.3)

Combining Like Terms

The expression $3x^2 + 2x + 1$ has three terms, $3x^2$, $2x$, and 1. A **term** is either a single number, a variable, or numbers and variables multiplied together. A term in an expression without a variable is called a **constant**, as 1 is above. For terms to be considered "like" terms, they must have the same variable and corresponding variables must have the same exponents. All constant terms are considered "like" terms.

like terms	unlike terms
$3x$ and $8x$	$9y$ and $10z$
$2x^2y$ and $3x^2y$	$3ab$ and $4ab^2$

In the example, $3x$ and $8x$ are like terms with numerical coefficients of 3 and 8. A **numerical coefficient** of a term is simply the number before its corresponding variable. When combining like terms, simply keep the variable the same and combine the numerical coefficients.

$4y + 10y = 14y$ $6b + 9b - 5b = 10b$
$10x - 3x = 7x$ $12x^2y - 10x^2y + 2x^2y = 4x^2y$

Identify the like terms in each problem.

1. $7c + 12c - 2$ **7c, 12c**
2. $19y - 10$ **none**
3. $12rt - 10r + 18t$ **none**
4. $5r - 10r + 8rs$ **5r, −10r**
5. $5d + 7d - 1$ **5d, 7d**
6. $q + 9 + 2q + 5q$ **q, 2q, 5q**

Simplify. If not possible, write *already simplified*.

7. $8m - 3m$ **5m**
8. $8y + 12y + 3y$ **23y**
9. $3s + 8s - 2$ **11s − 2**
10. $2 + 10k$ **already simplified**
11. $8q + 10q + 14$ **18q + 14**
12. $4 + 8x + 11y$ **already simplified**
13. $5a + 6a - 9a$ **2a**
14. $z + 8m + 4z - 4m$ **5z + 4m**
15. $5w + 2 + w$ **6w + 2**

Name_______________ (6.EE.A.2b, 6.EE.A.3)

Combining Like Terms

$4x + 5y + (^-18x) = ^-14x + 5y$

Combine like terms.

1. $3yz + 5yz =$ **8yz**
2. $3a + 5 + a =$ **4a + 5**
3. $5x - 5y - 8y + 8x =$ **13x − 13y**
4. $18x + 3x =$ **21x**
5. $5 - (^-4k) =$ **5 + 4k**
6. $7c - 12c =$ **−5c**
7. $13ab + (^-12ab) =$ **ab**
8. $^-12x + (^-4x) =$ **−16x**
9. $^-10n - (^-13n) =$ **3n**
10. $12b + (^-34b) =$ **−22b**
11. $4.7x - 5.9x =$ **−1.2x**
12. $4x^2 + (^-8y) + (^-3xy) + 5x^2 + 2xy =$ **$9x^2 - xy - 8y$**
13. $4x + 3y + (^-5y) + 3xy + y =$ **4x − y + 3xy**
14. $2x - y + 2x + 3xy =$ **4x + 3xy − y**
15. $5x + 7x =$ **12x**
16. $23x + 8 + 6x + 3y =$ **29x + 3y + 8**
17. $^-e + 8e =$ **7e**
18. $2xy + 5x + 6xy + 3xy + (^-3x) =$ **2x + 11xy**
19. $7s + 5x - 8s =$ **5x − s**
20. $4xy + 7xy + 6x^2y + 3xy^2 =$ **$11xy + 6x^2y + 3xy^2$**

Answer Key

Name______________________ (6.EE.A.2b, 6.EE.A.3)

Combining Like Terms

Combine like terms.

1. $^{-}n + 9n + 3 - 8 - 8n =$ **$^{-}5$**
2. $3(^{-}4x + 5y) - 3x(2 + 4y) =$ **$-18x + 15y - 12xy$**
3. $5 - 4y + x + 9y =$ **$5 + x + 5y$**
4. $^{-}2x + 3y - 5x - {}^{-}8y + 9y =$ **$^{-}7x + 20y$**
5. $6(a - b) - 5(2a + 4b) =$ **$^{-}4a - 26b$**
6. $7(x + 5y) + 3(x + 5y) + 5(3x + 8y) =$ **$25x + 90y$**
7. $12x + 6x + 9x - 3y + (^{-}7y) + y =$ **$27x - 9y$**
8. $^{-}21x + (^{-}2x) =$ **$^{-}23x$**
9. $4(x + 9y) - 2(2x + 4y) =$ **$28y$**
10. $4(x + 5y) + (5x + y) =$ **$9x + 21y$**
11. $6x + {}^{-}2y^2 + 4xy^2 + 3x^2 + 5xy^2 =$ **$3x^2 + 6x - 2y^2 + 9xy^2$**
12. $^{-}2(c - d) + (c - 3d) - 5(c - d) =$ **$^{-}6c + 4d$**
13. $3x + (^{-}3y) - (4x) + y =$ **$^{-}x - 2y$**
14. $^{-}3(4x + {}^{-}2y) - 2(x + 3y) - 2(2x + 6y) =$ **$-18x - 12y$**
15. $2b + 3(2b + 8a) - 3(8b + 2a) =$ **$18a + {}^{-}16b$**
16. $3[2(^{-}y^2 + y)\,^{-}3] - 3(2x + y) =$ **$^{-}6y^2 + 3y - 6x - 9$**
17. $2 \cdot 4x \cdot 3y - 4x \cdot 7y =$ **$^{-}4xy$**
18. $5(3a^2 - 2b^2) + 3a(a + 3b^2) =$ **$18a^2 + 9ab^2 - 10b^2$**
19. $3c + 4d + 2c + 5d - 4c =$ **$c + 9d$**
20. $4(x^2 + 3y^2) - y(x^2 + 5y) =$ **$4x^2 - xy^2 + 7y^2$**

Name______________________ (8.EE.C.7b, HSA-REI.B.3)

Solving One-Step Equations (Addition and Subtraction)

When solving equations for a given variable, use the addition or the subtraction property of equality. The addition property means that adding the same number to both sides of an equation will produce an equivalent equation.

$x - 3 = 5$
$x - 3 + 3 = 5 + 3$
$x = 8$

Solve for x by adding 3 to both sides of the equation.

When 3 is added to both sides, $^{-}3$ and 3 gives a result of 0, leaving x by itself. Therefore, $x = 8$ is the answer.

The subtraction property means that subtracting the same number from both sides of an equation will produce an equivalent equation.

$x + 3 = 5$
$x + 3 - 3 = 5 - 3$
$x = 2$

Solve for x by subtracting 3 from both sides of the equation.

When 3 is subtracted from both sides, $3 - 3$ gives a result of 0, leaving x by itself. Therefore, $x = 2$ is the answer.

Answers can be checked by substituting the value of x back into the equation. This is to make sure the value of x makes a true sentence when substituted into the original equation.

$x - 3 = 5, x = 8$
$8 - 3 = 5$
$5 = 5$

$x + 3 = 5, x = 2$
$2 + 3 = 5$
$5 = 5$

Solve each equation for x. Check your answers.

1. $x - 10 = 23$ **$x = 33$**
2. $6 + x = {}^{-}11$ **$x = {}^{-}17$**
3. $^{-}13 = {}^{-}6 + x$ **$x = {}^{-}7$**
4. $7 = 14 + x$ **$x = {}^{-}7$**
5. $8 = x + 9$ **$x = {}^{-}1$**
6. $x - 5 = {}^{-}5$ **$x = 0$**
7. $x + 3 = 12$ **$x = 9$**
8. $7 + x = 7$ **$x = 0$**
9. $7 + x = {}^{-}7$ **$x = {}^{-}14$**
10. $^{-}2 = x - 5$ **$x = 3$**
11. $x + 4 = {}^{-}15$ **$x = {}^{-}19$**
12. $9 = x + 12$ **$x = {}^{-}3$**

Name______________________ (8.EE.C.7b, HSA-REI.B.3)

Solving One-Step Equations (Addition and Subtraction)

$12 + x = {}^{-}24$
$12 - 12 + x = {}^{-}24 - 12$
$x = {}^{-}36$

Solve each equation for the given variable.

1. $^{-}13 + b = 31$ **$b = 44$**
2. $x - 17 = {}^{-}27$ **$x = {}^{-}10$**
3. $27 = v + (^{-}5)$ **$v = 32$**
4. $^{-}4 = x - 3$ **$x = {}^{-}1$**
5. $12 - (^{-}z) = 17$ **$z = 5$**
6. $^{-}200 = b + (^{-}73)$ **$b = {}^{-}127$**
7. $^{-}13 + x = 18$ **$x = 31$**
8. $^{-}w + (^{-}7) = {}^{-}56$ **$w = 49$**
9. $3 + x = 9$ **$x = 6$**
10. $z + 3.5 = 4.7$ **$z = 1.2$**
11. $12 + (^{-}g) = 10$ **$g = 2$**
12. $y - 12 = 15$ **$y = 27$**
13. $x + 2 = {}^{-}6$ **$x = {}^{-}8$**
14. $s - 5 = {}^{-}8$ **$s = {}^{-}3$**
15. $^{-}13 = n + (^{-}39)$ **$n = 26$**
16. $r = 4.4 + 3.9$ **$r = 8.3$**

Name______________________ (8.EE.C.7b, HSA-REI.B.3)

Solving One-Step Equations (Addition and Subtraction)

Solve each equation for the given variable.

1. $c + 23 = 41$ **$c = 18$**
2. $48 - x = 15$ **$x = 33$**
3. $a + 5.7 = 18.9$ **$a = 13.2$**
4. $^{-}29 - n = 6$ **$n = {}^{-}35$**
5. $40 - (^{-}g) = 38$ **$g = {}^{-}2$**
6. $c - 3 = 4.7$ **$c = 7.7$**
7. $n + \frac{3}{8} = \frac{5}{8}$ **$n = \frac{1}{4}$**
8. $^{-}1.9 + b = {}^{-}4.5$ **$b = {}^{-}2.6$**
9. $z + (^{-}14) = 13$ **$z = 27$**
10. $^{-}135 + r = {}^{-}26$ **$r = 109$**
11. $2\frac{1}{3} + w = 4\frac{2}{9}$ **$w = 1\frac{8}{9}$**
12. $^{-}0.6 - m = {}^{-}1.5$ **$m = 0.9$**
13. $^{-}d + (^{-}61) = 107$ **$d = -168$**
14. $s - \frac{7}{8} = -\frac{3}{4}$ **$s = \frac{1}{8}$**
15. $^{-}101 = g - (^{-}28)$ **$g = {}^{-}129$**
16. $241 + p = {}^{-}93$ **$p = {}^{-}334$**
17. $35.02 - q = 46.1$ **$q = {}^{-}11.08$**
18. $5\frac{3}{10} + (^{-}v) = 3.8$ **$v = 1.5$**
19. $^{-}74 = {}^{-}k - (^{-}91)$ **$k = 165$**
20. $^{-}8 + x = 5.62$ **$x = 13.62$**

Answer Key

Name____________________________ (8.EE.C.7b, HSA-REI.B.3)

Solving One-Step Equations (Multiplication and Division)

When solving equations for a given variable, use the multiplication or division property of equality. The multiplication property means that multiplying the same nonzero number by both sides of the equation will produce an equivalent equation.

$\frac{x}{7} = 3$ Multiply both sides of the equation by 7 to isolate x.

$(7)(\frac{x}{7}) = (7)(3)$ Check $\frac{21}{7} = 3$

$x = 21$ $3 = 3$

Therefore, the solution is 21.

The division property means that dividing both sides of the equation by the same nonzero number will produce an equivalent equation.

$^-5x = 20$ Divide both sides of the equation by $^-5$ to isolate x.

$\frac{^-5x}{^-5} = \frac{20}{^-5}$ Check $^-5(^-4) = 20$

$x = {}^-4$ $20 = 20$

Therefore, the solution is $^-4$.

Solve each equation for x. Check your answers.

1. $5x = 35$ **$x = 7$**
2. $18 = {}^-3x$ **$x = {}^-6$**
3. $^-7x = 49$ **$x = {}^-7$**
4. $-\frac{1}{3}x = 6$ **$x = {}^-18$**
5. $^-5x = {}^-20$ **$x = 4$**
6. $-\frac{5}{8}x = 10$ **$x = {}^-16$**
7. $\frac{1}{4}x = {}^-2$ **$x = {}^-8$**
8. $\frac{2}{3}x = 8$ **$x = 12$**
9. $4 = -\frac{x}{5}$ **$x = {}^-20$**
10. $^-4x = 48$ **$x = {}^-12$**
11. $\frac{x}{3} = {}^-5$ **$x = {}^-15$**
12. $^-6x = 24$ **$x = {}^-4$**

Name____________________________ (8.EE.C.7b, HSA-REI.B.3)

Solving One-Step Equations (Multiplication and Division)

$3x = 15$

$\frac{3x}{3} = \frac{15}{3}$

$x = 5$

$-\frac{3}{4y} = {}^-6$

$-\frac{4}{3} \cdot -\frac{3}{4y} = {}^-6 \cdot -\frac{4}{3}$

$\frac{1}{y} = 8$ so $y = \frac{1}{8}$

Solve each equation for the given variable.

1. $^-13h = 169$ **$h = {}^-13$**
2. $4b = {}^-36$ **$b = {}^-9$**
3. $10x = {}^-100$ **$x = {}^-10$**
4. $4c = 288$ **$c = 72$**
5. $7x = {}^-63$ **$x = {}^-9$**
6. $4y = {}^-48$ **$y = {}^-12$**
7. $6x = {}^-36$ **$x = {}^-6$**
8. $\frac{8}{k} = \frac{2}{5}$ **$k = 20$**
9. $^-(^-90) = {}^-45z$ **$z = {}^-2$**
10. $^-50 = 2x$ **$x = {}^-25$**
11. $\frac{2}{n} = \frac{1}{9}$ **$n = 18$**
12. $\frac{4}{x} = \frac{2}{9}$ **$x = 18$**
13. $\frac{x}{6} = \frac{6}{9}$ **$x = 4$**
14. $^-35c = 700$ **$c = {}^-20$**
15. $^-4x = {}^-20$ **$x = 5$**
16. $-\frac{x}{6} = \frac{2}{3}$ **$x = {}^-4$**

Name____________________________ (8.EE.C.7b, HSA-REI.B.3)

Solving One-Step Equations (Multiplication and Division)

Solve each equation for the given variable.

1. $12.8 = 4b$ **$b = 3.2$**
2. $-\frac{3}{4} = \frac{n}{16}$ **$n = {}^-12$**
3. $\frac{2}{3}x = {}^-10$ **$x = {}^-15$**
4. $^-(^-36) = {}^-0.25c$ **$c = {}^-144$**
5. $1.6r = 80$ **$r = 50$**
6. $-\frac{x}{8} = \frac{1}{4}$ **$x = {}^-2$**
7. $208x = {}^-4$ **$x = -\frac{1}{52}$**
8. $\frac{2}{k} = \frac{1}{8}$ **$k = 16$**
9. $^-4.9 = 7m$ **$m = {}^-0.7$**
10. $0.006w = {}^-0.54$ **$w = {}^-90$**
11. $\frac{720}{p} = 9$ **$p = 80$**
12. $-\frac{5h}{50} = {}^-0.1$ **$h = 1$**
13. $^-9s = 6$ **$s = -\frac{2}{3}$**
14. $15d = -\frac{1}{3}$ **$d = -\frac{1}{45}$**
15. $^-(^-27)d = 16.2$ **$d = 0.6$**
16. $3z = {}^-174$ **$z = {}^-58$**
17. $-\frac{4}{5j} = {}^-8$ **$j = \frac{1}{10}$**
18. $\frac{56}{63} = -\frac{x}{9}$ **$x = {}^-8$**

Name____________________________ (8.EE.C.7b, HSA-REI.B.3)

Solving Basic Multistep Equations

When solving equations for a given variable, sometimes you need to use more than one of the properties of equality.

$3x - 2 = 7$

$3x - 2 + 2 = 7 + 2$ Add 2 to both sides of the equation.

$\frac{3x}{3} = \frac{9}{3}$ Now, divide both sides of the equation by 3.

$x = 3$ Check $3(3) - 2 = 7$

$9 - 2 = 7$

$7 = 7$

Therefore, the solution is 3.

Here are some steps to follow when solving multi-step equations.

1. Simplify both sides of the equation (if possible).
2. Use the addition or subtraction property of equality to isolate terms containing the variable.
3. Use the multiplication or the division property of equality to further isolate the variable.
4. Check the solution.

Solve each equation for x. Check your answers.

1. $6x - 3 = 21$ **$x = 4$**
2. $^-6 + \frac{x}{4} = 1$ **$x = 28$**
3. $18 - 3x = {}^-12$ **$x = 10$**
4. $7 + 2x = {}^-13$ **$x = {}^-10$**
5. $^-4 = 7x + 8 - 8x$ **$x = 12$**
6. $13 = 9 - \frac{x}{5}$ **$x = {}^-20$**
7. $^-7 - x = {}^-5$ **$x = {}^-2$**
8. $5x + 9 - 4x = 12$ **$x = 3$**
9. $^-8x - 13 = 19$ **$x = {}^-4$**
10. $^-3 = {}^-5 - 2x$ **$x = {}^-1$**
11. $\frac{1}{2}x + 9 = 15$ **$x = 12$**
12. $^-7 = 3x - 15 - 7x$ **$x = {}^-2$**

Answer Key

Name________________________ (8.EE.C.7b, HSA-REI.B.3)

Solving Basic Multistep Equations

$$4x + 4 = 12$$
$$4x + 4 - 4 = 12 - 4$$
$$4x = 8$$
$$x = 2$$

Solve each equation for the given variable.

1. $7x - 12 = 2$ **$x = 2$**
2. $7a - 4 = 24$ **$a = 4$**
3. $4b - 7 = 37$ **$b = 11$**
4. $3c - 9 = 9$ **$c = 6$**
5. $8 - 9y = 35$ **$y = {}^-3$**
6. $8 - 12x = 32$ **$x = {}^-2$**
7. $1.3x + 5 = {}^-5.4$ **$x = {}^-8$**
8. $3(y + 4) + 5 = 35$ **$y = 6$**
9. $0 = 25x + 75$ **$x = {}^-3$**
10. $3 - \frac{1}{5}e = {}^-7$ **$e = 50$**
11. $5 - \frac{1}{2}x = {}^-9$ **$x = 28$**
12. $2x = 6 + ({}^-18)$ **$x = {}^-6$**
13. $7 - \frac{1}{9}k = 32$ **$k = {}^-225$**
14. $\frac{3}{12}w + 2 = 11$ **$w = 36$**
15. $\frac{2x}{5} + 3 = 9$ **$x = 15$**
16. $\frac{x}{3} - 8 = {}^-12$ **$x = {}^-12$**
17. $5(e + 5) = {}^-10$ **$e = {}^-7$**
18. $8 - \frac{1}{2}y = {}^-6$ **$y = 28$**

Name________________________ (8.EE.C.7b, HSA-REI.B.3)

Solving Basic Multistep Equations

Solve each equation for the given variable.

1. $5z - 8 = {}^-28$ **$z = {}^-4$**
2. $4k + 7 = {}^-9$ **$k = {}^-4$**
3. $13x + 7 = {}^-32$ **$x = {}^-3$**
4. $2x + 12 = 6$ **$x = {}^-3$**
5. $7.2 + 4x = 19.2$ **$x = 3$**
6. $2(w - 6) = 8$ **$w = 10$**
7. $7h + 1 = {}^-13$ **$h = {}^-2$**
8. $3(c - 2) = 15$ **$c = 7$**
9. $6x - 5 = {}^-41$ **$x = {}^-6$**
10. ${}^-3 + 2n = {}^-15$ **$n = {}^-6$**
11. $5e + ({}^-9) = 26$ **$e = 7$**
12. $\frac{m}{3} - 7 = {}^-10$ **$m = {}^-9$**
13. $6x - 2 = 34$ **$x = 6$**
14. ${}^-8(r - 2) = 40$ **$r = {}^-3$**
15. $5n - 8 = {}^-23$ **$n = {}^-3$**
16. $2 + (\frac{1}{5})x = {}^-7$ **$x = {}^-45$**
17. $5 - (\frac{1}{2})g = 12$ **$g = {}^-14$**
18. $3x - 4 = 14$ **$x = 6$**
19. ${}^-6 = \frac{3z}{4} + 12$ **$z = {}^-24$**
20. $2(f + 7) - 8 = 22$ **$f = 8$**
21. $4.7 = {}^-3.4m - 5.5$ **$m = {}^-3$**
22. $32 = \frac{4}{6}x - 34$ **$x = 99$**

Name________________________ (8.EE.C.7b, HSA-REI.B.3)

Solving Equations with Variables on Both Sides

Often, equations with a variable on both sides of the equation need to be solved. This requires one additional step of getting the variable on one side only.

$4x - 1 = 2x + 7$

$4x - 2x - 1 = 2x - 2x + 7$ Get the variable on one side of the equation.

$2x - 1 = 7$ Now, solve for x using the properties of equality.

$2x - 1 + 1 = 7 + 1$

$\frac{2x}{2} = \frac{8}{2}$ Check $4(4) - 1 = 2(4) + 7$

$x = 4$ $16 - 1 = 8 + 7$

$15 = 15$

Therefore, the solution is 4.

Combine the variables on the side of the equation with the greater variable coefficient, in order to avoid solving an equation with a variable which has a negative coefficient.

Solve each equation for x. Check your answers.

1. $9x - 12 = 3x$ **$x = 2$**
2. $8x - 12 = 15x - 4x$ **$x = {}^-4$**
3. $11 + 6x = 2x - 13$ **$x = {}^-6$**
4. ${}^-5x = 9 - 2x$ **$x = {}^-3$**
5. ${}^-8x - 10 = 4x + 14$ **$x = {}^-2$**
6. $10x - 5 = 21 - 3x$ **$x = 2$**
7. ${}^-12x = 14 - 5x$ **$x = {}^-2$**
8. $19 - 3x = 21 + x$ **$x = {}^-\frac{1}{2}$**
9. $4x + 12 = {}^-3x - 6 + 4x$ **$x = {}^-6$**
10. $14x + 5 = 8x - 1$ **$x = {}^-1$**
11. $7x - 3 = -4x - 25$ **$x = {}^-2$**
12. ${}^-9x + 5 = {}^-22 - 6x$ **$x = 9$**

Name________________________ (8.EE.C.7b, HSA-REI.B.3)

Solving Equations with Variables on Both Sides

$$6x - 7 = x + 33$$
$$6x - x - 7 = x - x + 33$$
$$5x - 7 + 7 = 33 + 7$$
$$5x = 40$$
$$x = 8$$

Solve each equation for the given variable.

1. $4x - 6 = 8 - 3x$ **$x = 2$**
2. $2(x + 3) = 12 - x$ **$x = 2$**
3. $5(5 - z) = 4(z - 5)$ **$z = 5$**
4. $4e + 6 = {}^-8 + 11e$ **$e = 2$**
5. ${}^-4(x - 6) = 2(7 - 7x)$ **$x = {}^-1$**
6. $2m - 9 = 8m - 27$ **$m = 3$**
7. $b + 9 = 6 + 2b$ **$b = 3$**
8. ${}^-9j + 3 = {}^-32 - 4j$ **$j = 7$**
9. $5(j - 4) = {}^-8 - j$ **$j = 2$**
10. $8({}^-9x + 4) = {}^-3(6x + 9) + 5$ **$x = 1$**
11. $3d - 5 = {}^-9 + 2d$ **$d = {}^-4$**
12. ${}^-2 + 6d = 34 - 3d$ **$d = 4$**
13. $3(k + 4) = {}^-3 - 2k$ **$k = {}^-3$**
14. $5(m - 3) = 27 - 2m$ **$m = 6$**
15. ${}^-j + 2 = {}^-14 - 5j$ **$j = {}^-4$**
16. ${}^-(x + 7) - 5 = 4(x + 3) - 6x$ **$x = 24$**
17. $5(x - 1) = 2x + 4(x - 1)$ **$x = {}^-1$**
18. ${}^-3(5k + 5) = 54 - 12k$ **$k = 23$**
19. $3e + 1 = 36 - 4e$ **$e = 5$**
20. ${}^-6r + 12 = {}^-2 + 8r$ **$r = 1$**

Answer Key

Name________________ (8.EE.C.7b, HSA-REI.B.3)

Solving Equations with Variables on Both Sides

Solve each equation for the given variable.

1. $7 - 6a = 6 - 7a$
   $a = ^-1$
2. $3c - 12 = 14 + 5c$
   $c = ^-13$
3. $3x - 3 = ^-3x + ^-3$
   $x = 0$
4. $2x - 7 = 3x + 4$
   $x = ^-11$
5. $9a + 5 = 3a - 1$
   $a = ^-1$
6. $8(x - 3) + 8 = 5x - 22$
   $x = ^-2$
7. $5d + 7 = 4d - 9$
   $d = -16$
8. $^-10w + 6 = ^-7w + ^-9$
   $w = 5$
9. $^-7c + 9 = c + 1$
   $c = 1$
10. $2n + 6 = 5n - 9$
    $n = 5$
11. $\frac{5}{2}x + 3 = \frac{1}{2}x + 15$
    $x = 6$
12. $5 + 3x = 7(x + 3)$
    $x = ^-4$
13. $12m - 9 = 4m + 15$
    $m = 3$
14. $2(f - 4) + 8 = 3f - 8$
    $f = 8$
15. $^-6 - (^-2n) = 3n - 6 + 5$
    $n = ^-5$
16. $4(2y - 4) = 5y + 2$
    $y = 6$
17. $2(r - 4) = 5[r + (^-7)]$
    $r = 9$
18. $6(x - 9) = 4(x - 5)$
    $x = 17$
19. $4(z + 5) - 3 = 6z - 13$
    $z = 15$
20. $4e - 19 = ^-3(e + 4)$
    $e = 1$
21. $4(^-6x - 2) = -22x$
    $x = ^-4$
22. $-\frac{1}{3}x - 5 = 7 - x$
    $x = 18$

© Carson-Dellosa • CD-104632 37

Name________________ (6.EE.A.2, 7.EE.B.3, 8.EE.C.7b)

Writing and Solving Equations

The sum of three times a number and 25 is 40. Find the number.

$$3x + 25 = 40$$
$$3x + 25 - 25 = 40 - 25$$
$$3x = 15$$
$$x = 5$$

The number is 5.

Write an equation for each word problem and solve it.

1. The difference of a number and $^-3$ is 8. Find the number.
   Equation **$x - (^-3) = 8$** Solution **$x = 5$**
2. Twice a number added to 9 is 15. Find the number.
   Equation **$2x + 9 = 15$** Solution **$x = 3$**
3. Twelve subtracted from 3 times a number is 15. Find the number.
   Equation **$3x - 12 = 15$** Solution **$x = 9$**
4. The sum of 4 times a number and 5 is $^-7$. Find the number.
   Equation **$4x + 5 = ^-7$** Solution **$x = ^-3$**
5. The product of a number and 5 is 60. Find the number.
   Equation **$5x = 60$** Solution **$x = 12$**
6. The difference of 5 times a number and 6 is 14. Find the number.
   Equation **$5x - 6 = 14$** Solution **$x = 4$**
7. The sum of a number and $^-6$ is 10. Find the number.
   Equation **$x + (^-6) = 10$** Solution **$x = 16$**
8. The quotient of a number and 4 is $^-12$. Find the number.
   Equation **$\frac{x}{4} = ^-12$** Solution **$x = ^-48$**

38 © Carson-Dellosa • CD-104632

Name________________ (6.EE.A.2, 7.EE.B.3, 8.EE.C.7b)

Writing and Solving Equations

Write an equation for each word problem and solve it.

1. Six times the difference of a number and 9 is 54. Find the number.
   Equation **$6(x - 9) = 54$** Solution **$x = 18$**
2. The sum of 8 times a number and 3 is 59. Find the number.
   Equation **$8x + 3 = 59$** Solution **$x = 7$**
3. The sum of 5 times a number and $^-11$ is $^-16$. Find the number.
   Equation **$5x + (^-11) = ^-16$** Solution **$x = ^-1$**
4. Twelve times the sum of a number and $^-8$ is 48. Find the number.
   Equation **$12[x + (^-8)] = 48$** Solution **$x = 12$**
5. The sum of 5 times a number and 2 is $^-13$. Find the number.
   Equation **$5x + 2 = ^-13$** Solution **$x = ^-3$**
6. The sum of 7 times a number and 11 is 81. Find the number.
   Equation **$7x + 11 = 81$** Solution **$x = 10$**
7. Three times the sum of a number and $^-2$ is $^-15$. Find the number.
   Equation **$3[x + (-2)] = ^-15$** Solution **$x = ^-3$**
8. Five times the sum of a number and 2 is 35. Find the number.
   Equation **$5(x + 2) = 35$** Solution **$x = 5$**

© Carson-Dellosa • CD-104632 39

Name________________ (6.EE.A.2, 7.EE.B.3, 8.EE.C.7b)

Writing and Solving Equations

Write an equation for each word problem and solve it.

1. A number increased by 10 is 38. Find the number.
   Equation **$x + 10 = 38$** Solution **$x = 28$**
2. A number times 8 is $^-96$. Find the number.
   Equation **$8x = ^-96$** Solution **$x = ^-12$**
3. Thirty-two is 7 less than 3 times a number. Find the number.
   Equation **$32 = 3x - 7$** Solution **$x = 13$**
4. A number multiplied by 4 and increased by 7 is $^-25$. Find the number.
   Equation **$^-25 = 4x + 7$** Solution **$x = ^-8$**
5. A number decreased by 12 is the same as 3 times the number. Find the number.
   Equation **$x - 12 = 3x$** Solution **$x = ^-6$**
6. Four times the sum of a number and 7 is 44 less than the number. Find the number.
   Equation **$4(x + 7) = x - 44$** Solution **$x = ^-24$**
7. A number increased by 16 is the same as 8 times the sum of the number and 9. Find the number.
   Equation **$16 + x = 8(x + 9)$** Solution **$x = ^-8$**
8. Four times a number is decreased by 9 and then increased by 12. The result is 5 less than 2 times the number. Find the number.
   Equation **$4x - 9 + 12 = 2x - 5$** Solution **$x = ^-4$**

40 © Carson-Dellosa • CD-104632

Answer Key

Name ______________________ (6.EE.B.8, HSA-REI.B.3)

Solving Inequalities

$-10n + 5 \le 55$
$-10n + 5 - 5 \le 55 - 5$
$-10n \le 50$
$n \ge -5$

Solve each inequality and graph its solution set.

1. $-4(3d + 2) \le 4$
$d \ge -1$

2. $10 - 5x - 20 \ge -20$
$x \le 2$

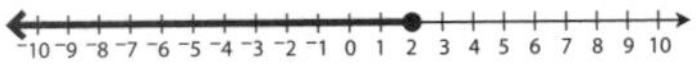

3. $-15 > 4y - 7 - 3y - 4$
$y < -4$

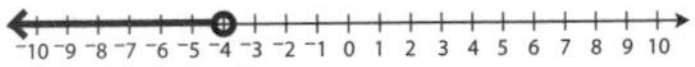

4. $4x - 7 < 9$
$x < 4$

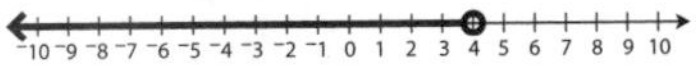

5. $6n - 3 > 33$
$n > 6$

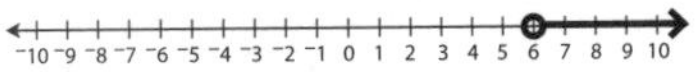

6. $3(3c - 4) \le 15$
$c \le 3$

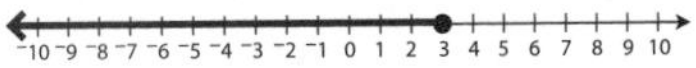

7. $5z - 1 > 9$
$z > 2$

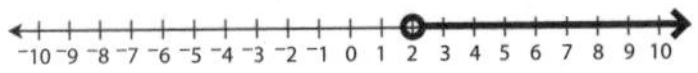

8. $5 > 4x - 11$
$x < 4$

Name ______________________ (6.EE.B.8, HSA-REI.B.3)

Solving Inequalities

$-10x > 4x - 42$
$-10x + 10x > 4x + 10x - 42$
$0 > 14x - 42$
$0 + 42 > 14x - 42 + 42$
$42 > 14x$
$x < 3$

Solve each inequality and graph its solution set.

1. $-2a < 10 + 3a$
$a > -2$

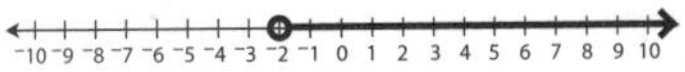

2. $4w > 2w + 8$
$w > 4$

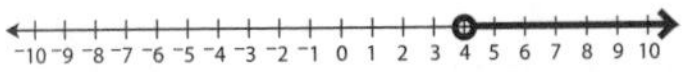

3. $2(k + 4) \le 3(2k - 4)$
$k \ge 5$

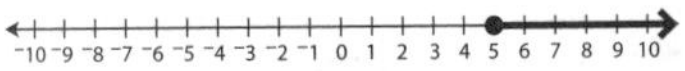

4. $5n + 3 \ge -12$
$n \ge -3$

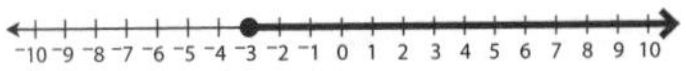

5. $5c + 2 < 2c + (-7)$
$c < -3$

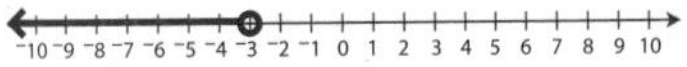

6. $5x - 20 > 2x + 1$
$x > 7$

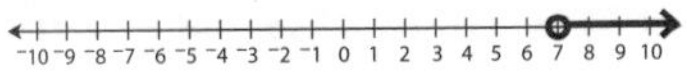

7. $3(s - 4) \ge 4s - 12$
$s \le 0$

8. $-9 - e > 3e + 11$
$e < -5$

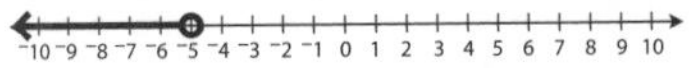

Name ______________________ (6.EE.B.8, HSA-REI.B.3)

Solving Inequalities

Solve each inequality and graph its solution set.

1. $15e - 3 \le 20e + 17$
$e \ge -4$

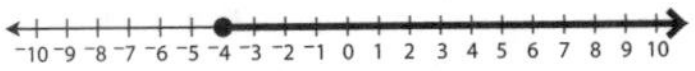

2. $6x - 4 > 2(x - 6)$
$x > -2$

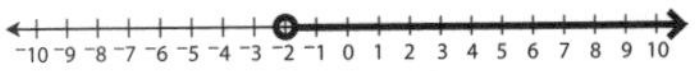

3. $7c - 8 \ge 6$
$c \ge 2$

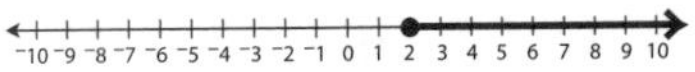

4. $5x + (-3) \ge 2(3 + x)$
$x \ge 3$

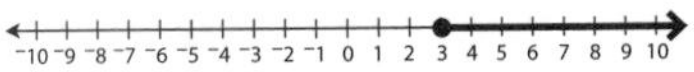

5. $-8 < 2(2 + 3r)$
$r > -2$

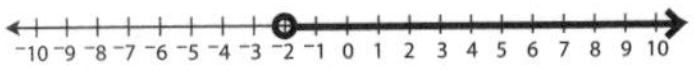

6. $6d < 3d - 18$
$d < -6$

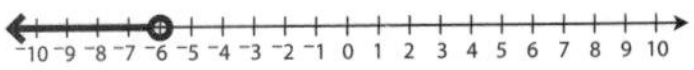

7. $7m + 9 \le 6(m + 3)$
$m \le 9$

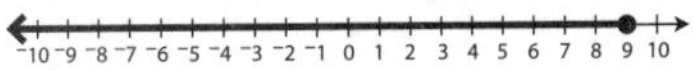

8. $3(2x + 4) \ge 7x + 8$
$x \le 4$

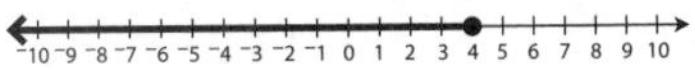

9. $-4.2 > 0.6y$
$y < -7$

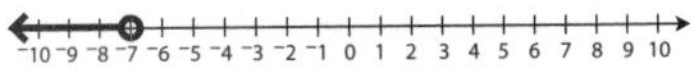

10. $\frac{a}{4} + 3 \le 5$
$a \le 8$

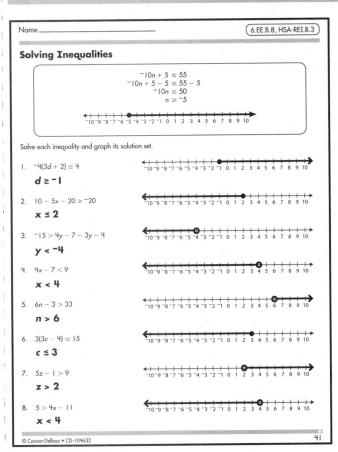

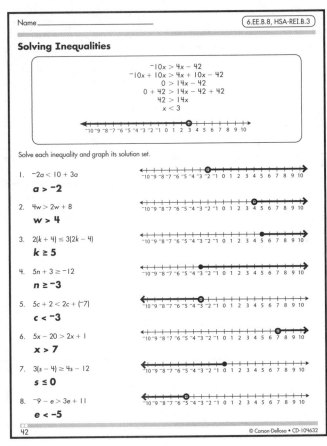

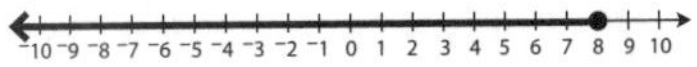

Name ______________________ (8.EE.A.1)

Multiplication and Powers of Exponents

When multiplying exponents, it is important to remember the following properties:

1. When multiplying powers having the same base, add the exponents, keeping the same base. (Remember: in a^3, a is the base, 3 is the exponent, and a^3 is the power.) For example, $x^3 \bullet x^5 = x^{3+5} = x^8$.
2. When finding a power of a power, multiply the exponents. For example, $(x^3)^2 = x^6$.
3. When finding the power of a product, find the power of each factor and multiply. For example, $(x \bullet y)^2 = x^2 \bullet y^2$.

Simplify $(5x^3)^2(xy)^3$
$(5^2x^6)(x^3y^3)$
$= 25x^9y^3$

Simplify $(2x^4)^3(-x^2)^3$
$(2^3x^{12})(-x^6)$
$= -8x^{18}$

1. What do you do with the exponents when multiplying powers that have the same base?
Add the exponents.
2. Label the base, the exponent, and the power in x^3.
x: base, 3: exponent, x^3: power
3. Explain what you are to do when finding the power of a product.
Find the power of each factor and multiply.

Simplify each expression.

4. $3x \bullet x^2$
$3x^3$
5. $(-6xy)^2(x^2y)^3$
$36x^8y^5$
6. $(8xy)^2$
$64x^2y^2$
7. $-4x^4 \bullet x^3$
$-4x^7$
8. $(-3x^2y)^3$
$-27x^6y^3$
9. $(4c^2)(-5c^7)$
$-20c^9$
10. $(-x^2)(-x)^2$
x^4
11. $-xy(-xy)^2$
$-x^3y^3$
12. $(3x^3)(5x^5)$
$15x^8$
13. $(x^3y^3)^3$
x^9y^9

Answer Key

Name______________________ (8.EE.A.1)

Multiplication and Powers of Exponents

Rule: $(x^a)^b = x^{ab}$ Example: $(x^2y^3)^3 = x^6y^9$

Rule: $x^a \cdot x^b = x^{a+b}$ Example: $x^3 \cdot x^5 = x^8$

Simplify each expression.

1. $c \cdot c^2 \cdot c^3 =$ $\mathbf{c^6}$
2. $e \cdot e^2 \cdot e^3 \cdot e^4 \cdot e^5 =$ $\mathbf{e^{15}}$
3. $a^3 \cdot a^4 \cdot a^7 \cdot a =$ $\mathbf{a^{15}}$
4. $(3xy^2)(2x^2y^3) =$ $\mathbf{6x^3y^5}$
5. $(2a^2b)(4ab^2) =$ $\mathbf{8a^3b^3}$
6. $(5f)(^-3f^3)(2f)^2 =$ $\mathbf{^-60f^6}$
7. $(m^2n)^3(4mn^2)(mn) =$ $\mathbf{4m^8n^6}$
8. $(4k^2)(^-3k)(3k^5) =$ $\mathbf{^-36k^8}$
9. $(^-2c^4)(4cd)(^-cd^2) =$ $\mathbf{8c^6d^3}$
10. $(3x^3)^2(3x^4)(^-3x^2)^3 =$ $\mathbf{^-729x^{16}}$
11. $(^-1)(x)(^-x^2)(x^3)(^-x^2) =$ $\mathbf{^-x^8}$
12. $(3x^2)(^-3x^5) =$ $\mathbf{^-9x^7}$
13. $(c^2h)^2(3ch^3)(2c^3h^4)^2 =$ $\mathbf{12c^{11}h^{13}}$
14. $(^-4p^3)(^-4p^6)(^-2p^9) =$ $\mathbf{^-32p^{18}}$
15. $(12c^3)(2g^3)(4ch)^3$ $\mathbf{1{,}536c^6g^3h^3}$
16. $(4x^2y^3)^3(x^3y)(^-x^2y^2) =$ $\mathbf{^-64x^{11}y^{12}}$
17. $(^-4f^3)^4(^-3m^3)^5 =$ $\mathbf{62{,}208f^{12}m^{15}}$
18. $(2c^2d^2)^4(^-5cd^4) =$ $\mathbf{^-80c^9d^{12}}$
19. $(3x)^2(2x^3y^6)(^-5x^6y^2) =$ $\mathbf{^-90x^{11}y^8}$
20. $(3x)(^-4y^2)(6x^3y) =$ $\mathbf{^-72x^4y^3}$

© Carson-Dellosa • CD-104632 45

Name______________________ (8.EE.A.1)

Multiplication and Powers of Exponents

Simplify each expression.

1. $(^-4xy^3)^3 =$ $\mathbf{-64x^3y^9}$
2. $(x^2y^3)(x^3y) =$ $\mathbf{x^5y^4}$
3. $(^-6x^4y^6)^3 =$ $\mathbf{^-216x^{12}y^{18}}$
4. $(5x^2y^4)^3 =$ $\mathbf{125x^6y^{12}}$
5. $(6x^5y^4)(5x^5y^4) =$ $\mathbf{30x^{10}y^8}$
6. $(2x)^4(x^3y)^2 =$ $\mathbf{16x^{10}y^2}$
7. $(4x^3y^2)^3(^-2x^2y^4) =$ $\mathbf{^-128x^{11}y^{10}}$
8. $(3x^2y)(^-8xy^4) =$ $\mathbf{^-24x^3y^5}$
9. $(^-2x^2y)^4 =$ $\mathbf{16x^8y^4}$
10. $(x^2y^3)^2(x^3y^2)^4 =$ $\mathbf{x^{16}y^{14}}$
11. $(^-4x^3y^3)^4(^-8x^3)^2 =$ $\mathbf{16{,}384x^{18}y^{12}}$
12. $(3xy^3)(^-4x^2y^4)^2(xy^3) =$ $\mathbf{48x^6y^{14}}$
13. $(^-3x^3y)^3 =$ $\mathbf{^-27x^9y^3}$
14. $(^-2x^4y^5)^3 =$ $\mathbf{^-8x^{12}y^{15}}$
15. $(3x^2y^3)^4(x^3y)(7xy^3) =$ $\mathbf{567x^{12}y^{16}}$
16. $(6x^2y^3)(4xy^2)^3(3x^2y)^2 =$ $\mathbf{3{,}456x^9y^{11}}$
17. $(^-2x^3y^3z)^4(2xyz^4)^2 =$ $\mathbf{64x^{14}y^{14}z^{12}}$
18. $(5xy^3)(^-5xy^2)^5 =$ $\mathbf{^-15{,}625x^6y^{13}}$
19. $(^-3x^2y^3)^2 =$ $\mathbf{9x^4y^6}$
20. $(4z^4)^2(2x^2y)(^-3xy^3z^5) =$ $\mathbf{^-96x^3y^4z^{13}}$

46 © Carson-Dellosa • CD-104632

Name______________________ (8.EE.A.1)

Division of Exponents

When dividing exponents, it is important to remember the following properties:

1. When dividing powers that have the same base, subtract the exponents.

   For example, $\frac{x^4}{x^2} = x^{4-2} = x^2$, where x cannot be equal to 0.

2. When finding a power of a quotient, find the power of the numerator and the power of the denominator and divide.

   For example, $(\frac{x}{y})^3 = \frac{x^3}{y^3}$, where y cannot be equal to 0.

Simplify: $\frac{6^8}{6^6}$ $= 6^{8-6}$ $6^2 = 36$

$(\frac{3}{4})^{-2}$ $= \frac{3^{-2}}{4^{-2}}$ $= \frac{4^2}{3^2} = \frac{16}{9}$

1. Explain what you do with the exponents when dividing powers that have the same base.

   Subtract the exponents.

Evaluate each expression.

2. $\frac{5^6}{5^3}$ **125**
3. $\frac{(^-3)^2}{3^2}$ **1**
4. $\frac{3^5}{3^2}$ **27**
5. $\frac{5^4 \cdot 5}{5^7}$ $\mathbf{\frac{1}{25}}$
6. $\frac{8^6}{8^3}$ **512**
7. $\frac{7^3}{7}$ **49**
8. $\frac{4^8}{4^8}$ **1**
9. $\frac{6^4 \cdot 6^3}{6^5}$ **36**
10. $\frac{4^6}{4^2}$ **256**
11. $\frac{10^5}{10^3}$ **100**

Simplify each expression.

12. $\frac{x^5}{x^3}$ $\mathbf{x^2}$
13. $x^7 \cdot \frac{1}{x^5}$ $\mathbf{x^2}$
14. $\frac{18x^4y^4}{6x^2y^4}$ $\mathbf{3x^2}$
15. $\frac{4x^4y^4}{4x^2y}$ $\mathbf{x^2y^3}$
16. $\frac{6x^2y^4}{3y^2}$ $\mathbf{2x^2y^2}$
17. $x^4 \cdot \frac{1}{x^2}$ $\mathbf{x^2}$

© Carson-Dellosa • CD-104632 47

Name______________________ (8.EE.A.1)

Division of Exponents

Rule: $\frac{x^a}{x^b} = x^{a-b}$ Example: $\frac{x^6}{x^4} = x^{6-4} = x^2$ $\frac{x^3}{x^{-2}} = x^{3-(^-2)} = x^5$

Simplify each expression.

1. $\frac{^-12m^5}{6m} =$ $\mathbf{^-2m^4}$
2. $\frac{x^3}{x^2} =$ $\mathbf{x}$
3. $\frac{9a^3b^5}{^-3ab^2} =$ $\mathbf{^-3a^2b^3}$
4. $\frac{16c^3}{^-4c^2} =$ $\mathbf{^-4c}$
5. $\frac{d^3}{d^2} =$ $\mathbf{d}$
6. $\frac{^-3p^8}{6p^2} =$ $\mathbf{\frac{(^-p^6)}{2}}$
7. $\frac{^-54c^2d^4}{^-8cd} =$ $\mathbf{\frac{27cd^3}{4}}$
8. $\frac{49r^{13}}{^-7r^8} =$ $\mathbf{^-7r^5}$
9. $\frac{45k^7r^3}{^-3k^5} =$ $\mathbf{^-15k^2r^3}$
10. $\frac{^-14c^{15}d^3}{^-2c^9d} =$ $\mathbf{7c^6d^2}$
11. $\frac{(5k)(^-8k^5)}{10k^3} =$ $\mathbf{^-4k^3}$
12. $\frac{24x^2y}{^-4x^2} =$ $\mathbf{^-6y}$
13. $\frac{4x^2y^3z^4}{2xy^2z^3} =$ $\mathbf{2xyz}$
14. $\frac{9a^{11}}{a^3} =$ $\mathbf{9a^8}$
15. $\frac{22y^5z^8}{2yz^7} =$ $\mathbf{11y^4z}$
16. $\frac{b^{14}c^9}{b^5c^4} =$ $\mathbf{b^9c^5}$

48 © Carson-Dellosa • CD-104632

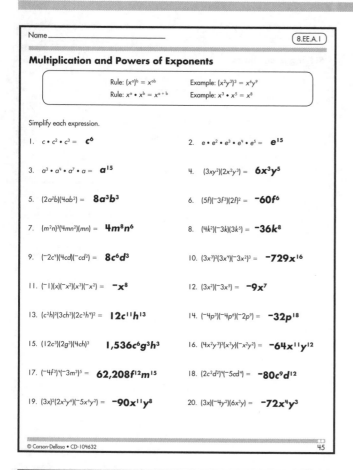

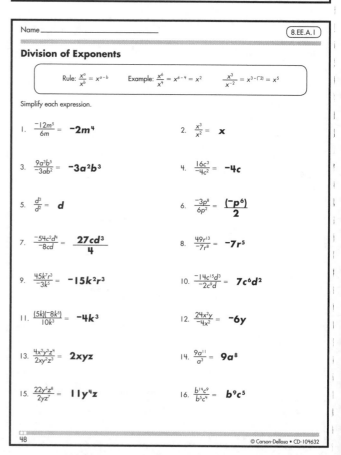

Answer Key

Name________________ (8.EE.A.1)

Division of Exponents

Simplify.

1. $\frac{a^5}{a^3} = \mathbf{a^2}$
2. $\frac{a^5b^2}{2a^2} = \mathbf{\frac{a^3b^2}{2}}$
3. $\frac{13m^6n^7}{39m^3n^5} = \mathbf{\frac{m^3n^2}{3}}$
4. $\frac{9x^8y^7z^8}{18x^5y^5z^4} = \mathbf{\frac{x^3y^2z^4}{2}}$
5. $\frac{-2c^2d^3}{-8cd^2} = \mathbf{\frac{cd}{4}}$
6. $\frac{10a^6b^8}{40a^2b^2} = \mathbf{\frac{a^4b^6}{4}}$
7. $\frac{18a^6b^2c^6}{36a^4bc^2} = \mathbf{\frac{a^2bc^4}{2}}$
8. $\frac{5x^3y^2z^2}{5x^2yz} = \mathbf{xyz}$
9. $\frac{45x^9y^{10}z^5}{51x^9y^8z^3} = \mathbf{\frac{15y^2z^2}{17}}$
10. $\frac{16x^2y^4}{4x^2y^3} = \mathbf{4y}$
11. $\frac{18x^6y^3z^4}{12x^3y^2z^3} = \mathbf{\frac{3x^3yz}{2}}$
12. $\frac{72x^5y^5z^6}{8x^4yz^3} = \mathbf{9xy^4z^3}$
13. $\frac{18a^9b^3}{12a^2b^2} = \mathbf{\frac{3a^7b}{2}}$
14. $\frac{44x^8y^2}{11x^7y} = \mathbf{4xy}$
15. $\frac{(6x^3)(3x^8)}{-12x^{10}} = \mathbf{\frac{-3x}{2}}$
16. $\frac{21k^9}{(3k)(7k^4)} = \mathbf{k^4}$
17. $\frac{(110c^3)(-c^9)}{11c^5} = \mathbf{-10c^7}$
18. $\frac{(3xy)(4x^2y)}{-6xy^2} = \mathbf{-2x^2}$

Name________________ (8.EE.A.1)

Negative and Zero Exponents

Given a nonzero number a and a positive integer n, the following definitions of negative exponents and zero exponents are stated.

1. For a negative exponent: the expression a^{-n} is the reciprocal of a^n. This is written: $a^{-n} = \frac{1}{a^n}$, where $a \neq 0$.
2. For a zero exponent: any nonzero number raised to the 0 power will have an answer of 1. This is written: $a^0 = 1$, where $a \neq 0$.

Evaluate: 5^{-3}

$= \frac{1}{5^3}$

$= \frac{1}{125}$

Simplify by rewriting with positive exponents. $2x^{-3}y^2z^{-4}$

$= 2 \bullet \frac{1}{x^3} \bullet y^2 \bullet \frac{1}{z^4}$

$= \frac{2y^2}{x^3z^4}$

Note: When rewriting expressions to positive exponent form, simply move the factors from the denominator to the numerator or vice versa, leaving out the in-between step.

1. Any number raised to the zero power has what value?

   1

Evaluate each expression.

2. 4^{-2} $\mathbf{\frac{1}{16}}$
3. $7^3 \bullet 7^{-3}$ **1**
4. $3 \bullet 3^{-1}$ **1**
5. $-5^0 \bullet \frac{1}{3^{-3}}$ $\mathbf{-27}$
6. $(6^2)^{-2}$ $\mathbf{\frac{1}{1296}}$
7. $(-2^{-3})^{-1}$ $\mathbf{-8}$

Rewrite each expression with positive exponents.

8. x^{-8} $\mathbf{\frac{1}{x^8}}$
9. $\frac{1}{3x^{-2}}$ $\mathbf{\frac{x^2}{3}}$
10. $(-3)^0x^{-2}$ $\mathbf{\frac{1}{x^2}}$
11. x^{-10} $\mathbf{\frac{1}{x^{10}}}$
12. $x^{-3}y^4$ $\mathbf{\frac{y^4}{x^3}}$
13. $\frac{6}{x^{-2}}$ $\mathbf{6x^2}$

Name________________ (8.EE.A.1)

Negative and Zero Exponents

Rule: $x^{-a} = \frac{1}{x^a}$

Example: $4^{-2} = \frac{1}{16}$

$4^{-2} = \frac{1}{4^2} = \frac{1}{16}$

Example: $4x^{-2} = \frac{4}{x^2}$

Example: $(2x)^{-3} = \frac{1}{8x^3}$

$\frac{1}{(2x)^3} = \frac{1}{8x^3}$

Rule: $x^0 = 1$

Simplify.

1. $4cd^{-5}$ $\mathbf{\frac{4c}{d^5}}$
2. 3^0 **1**
3. $3a^4b^{-3}$ $\mathbf{\frac{3a^4}{b^3}}$
4. 4^{-5} $\mathbf{\frac{1}{1024}}$
5. $(-2)^{-2}$ $\mathbf{\frac{1}{4}}$
6. $(3xy)^{-1}$ $\mathbf{\frac{1}{3xy}}$
7. $(3x)^{-3}$ $\mathbf{\frac{1}{27x^3}}$
8. $7x^{-3}$ $\mathbf{\frac{7}{x^3}}$
9. $-2x^{-3}$ $\mathbf{\frac{-2}{x^3}}$
10. $(6y^2)^{-2}$ $\mathbf{\frac{1}{36y^4}}$
11. $(\frac{4}{5})^{-2}$ $\mathbf{\frac{25}{16}}$
12. $4m^3n^{-5}$ $\mathbf{\frac{4m^3}{n^5}}$
13. $(-11x^3y)^{-2}$ $\mathbf{\frac{1}{121x^6y^2}}$
14. $(c^0d)^{-2}$ $\mathbf{\frac{1}{d^2}}$
15. $(\frac{x^2}{y^3})^{-2}$ $\mathbf{\frac{y^6}{x^4}}$
16. $(\frac{2}{3})^{-1}$ $\mathbf{\frac{3}{2}}$
17. b^0 **1**
18. c^{-7} $\mathbf{\frac{1}{c^7}}$

Name________________ (8.EE.A.1)

Negative and Zero Exponents

Simplify.

1. $6x^{-3}$ $\mathbf{\frac{6}{x^3}}$
2. $4x^{-5}$ $\mathbf{\frac{4}{x^5}}$
3. $\frac{1}{5x^{-4}}$ $\mathbf{\frac{x^4}{5}}$
4. $4x^{-4}y^{-2}$ $\mathbf{\frac{4}{x^4y^2}}$
5. $\frac{1}{(2x)^{-3}}$ $\mathbf{8x^3}$
6. $(3x^{-2})^2$ $\mathbf{\frac{9}{x^4}}$
7. $14x^{-8}y$ $\mathbf{\frac{14y}{x^8}}$
8. $(-5x^3)^{-2}$ $\mathbf{\frac{1}{25x^6}}$
9. $(-3x^{-2}y^5z^3)^{-2}$ $\mathbf{\frac{x^4}{9y^{10}z^6}}$
10. $8x^{-2}y^{-4}$ $\mathbf{\frac{8}{x^2y^4}}$
11. $2x^0y^{-5}z^2$ $\mathbf{\frac{2z^2}{y^5}}$
12. $(4x^3y^{-4})^2$ $\mathbf{\frac{16x^6}{y^8}}$
13. $(\frac{x^3}{2y^{-2}})^{-3}$ $\mathbf{\frac{8}{x^9y^6}}$
14. $4m^{-5}n^{-1}$ $\mathbf{\frac{4}{m^5n}}$
15. $(\frac{x^2}{y^3})^{-2}$ $\mathbf{\frac{y^6}{x^4}}$
16. $(-14c^2d)^{-1}$ $\mathbf{\frac{1}{-14c^2d}}$
17. $9xy^{-4}$ $\mathbf{\frac{9x}{y^4}}$
18. $(-6s^3p^2)^{-3}$ $\mathbf{\frac{1}{-216s^9p^6}}$

Answer Key

Name________________ (7.EE.A.1, HSA-APR.A.1)

Adding and Subtracting Polynomials

When adding polynomials, the following two methods can be used.

1. The horizontal method:

   Add $(3x^4 + 8x^2 - 2)$ and $({}^-4x^2 + 7x^4 + 5)$.
   First group like terms in descending order. $(3x^4 + 7x^4) + (8x^2 - 4x^2) + ({}^-2 + 5)$
   Simplify. $10x^4 + 4x^2 + 3$

2. The vertical method:

   Add $(4x^7 + 6x^4 - 10 + 2x^5)$ and $(10x^5 - 5x^7 - 3x^4 + 12)$.
   Group like terms in columns in descending order.

$$\begin{array}{r} 4x^7 + 2x^5 + 6x^4 - 10 \\ + \quad {}^-5x^7 + 10x^5 - 3x^4 + 12 \\ \hline {}^-x^7 + 12x^5 + 3x^4 + 2 \end{array}$$

   Simplify.

When subtracting polynomials, change the sign of each term in the second polynomial and then choose from one of the methods above to simplify.

Subtract. $(10x^7 - 4x^4 + 7x^2 + 2) - (12x^4 + 6x^7 + 8 - 5x^2)$

Change sign of each term in second polynomial. $10x^7 - 4x^4 + 7x^2 + 2 - 12x^4 - 6x^7 - 8 + 5x^2$

Choosing method 1 above, group like terms in descending order.

$(10x^7 - 6x^7) + ({}^-4x^4 - 12x^4) + (7x^2 + 5x^2) + (2 - 8)$

Simplify. $4x^7 - 16x^4 + 12x^2 - 6$

1. What is important to remember to do when subtracting polynomials?

Change the sign of each term in the second polynomial.

Simplify.

2. $(7x - 5) - (3x + 7)$
   $4x - 12$
3. $(2x^2 - 3x + 1) - (5x^2 - 3x + 10)$
   $-3x^2 - 9$
4. $(6x + 7) + (8x - 3)$
   $14x + 4$
5. $(9x^2 + 3x - 5) - (3x^2 + 4x - 10)$
   $6x^2 - x + 5$
6. $(8x^4 - 9x^3 + 2x) - (6x^4 + 7x^3 - 3x)$
   $2x^4 - 16x^3 + 5x$
7. $(x^3 - x^2 + 2) - (x^3 + x^2 + 5)$
   $-2x^2 - 3$

Name________________ (7.EE.A.1, HSA-APR.A.1)

Adding and Subtracting Polynomials

$(x^2 + 4x + 2) - (2x^2 + 7x - 6) = {}^-x^2 - 3x + 8$

Add or subtract the polynomials by combining like terms.

1. $(4y^2 - 9y) - ({}^-5y^2 + 8y - 8) =$ **$9y^2 - 17y + 8$**
2. $(6x^2 + 2x + 6) - (4x^2 - 2x + 3) + ({}^-5x^2 + 5x + 6) =$ **$-3x^2 + 9x + 9$**
3. $({}^-2x^3 + 3x^2 + 9) + ({}^-8x^3 - 2x^2 + {}^-4x) =$ **$-10x^3 + x^2 - 4x + 9$**
4. $(2x^2 - 9x - 8) - (2x^3 - 7x^2 + {}^-2) =$ **$-2x^3 + 9x^2 - 9x - 6$**
5. $(4x^3 - 2x^2 - 12) + (6x^2 + 3x + 8) =$ **$4x^3 + 4x^2 + 3x - 4$**
6. $(3x^4 - 3x + 1) - (4x^3 - 4x - 8) =$ **$3x^4 - 4x^3 + x + 9$**
7. $(7x^2 - x - 5) - (3x^2 - 3x + 5) =$ **$4x^2 + 2x - 10$**
8. $(x^3 - x^2 + 3) - (3x^3 - x^2 + 7) =$ **$-2x^3 - 4$**
9. $({}^-2x^2 + 4x - 12) + (5x^2 - 5x) =$ **$3x^2 - x - 12$**
10. $(4x^3 - 5x^2 - 9) - (6x^3 - 5x - 4) - (5x^3 - 4x^2 - 10) =$ **$-7x^3 - x^2 + 5x + 5$**

Name________________ (7.EE.A.1, HSA-APR.A.1)

Adding and Subtracting Polynomials

Add or subtract the polynomials by combining like terms.

1. $(x^3 - 7x^4 + 2x^2 - 3x) + (4x^4 - 8x^3 + 5x) =$ **$-3x^4 - 7x^3 + 2x^2 + 2x$**
2. $(6x^2 + 9x + 4) - ({}^-7x^2 + 2x - 1) =$ **$13x^2 + 7x + 5$**
3. $({}^-8x^2 + 2) + (7x - 5) =$ **$-8x^2 + 7x - 3$**
4. $(8x + 6) - (10x + 16) + (4x - 6) =$ **$2x - 16$**
5. $(6x^2 + 7x - 9) + (3x + 8) =$ **$6x^2 + 10x - 1$**
6. $(4x^2 + 2) + (3x^2 - 4x + 6) - (5x^2 + 10) =$ **$2x^2 - 4x - 2$**
7. $(2x^2 + 3x + 2) - (6x^3 - 3x^2 + 8) - ({}^-2x^3 + 9x^2 + 7) =$ **$-4x^3 - 4x^2 + 3x - 13$**
8. $({}^-3x^2 - 4x^3 - 1) - (2x^3 - 7x - 9) - (2x^3 - 2x^2 - 3) =$ **$-8x^3 - x^2 + 7x + 11$**
9. $({}^-6x^2 - 3x^3 + 4) + ({}^-7x^3 + 2x + 4) - ({}^-3x^3 + 5x^2 + 2) =$ **$-7x^3 - 11x^2 + 2x + 6$**
10. $(4x^2 + 6x + 3) + (3x^2 - 3x - 2) + ({}^-4x^2 + 3x - 9) =$ **$3x^2 + 6x - 8$**
11. $(9x^2 - 7x + {}^-4) + (3x^3 - 4x + {}^-5) + ({}^-4x^2 - 2x - 5) =$ **$3x^3 + 5x^2 - 13x - 14$**
12. $(3x^3 - 4x + 3) + ({}^-3x^3 + 5x - 8) =$ **$x - 5$**
13. $({}^-5x^5 + 7x^3 - 4x + 2) + (7x^5 - 8x^4 + 9x^3) =$ **$2x^5 - 8x^4 + 16x^3 - 4x + 2$**
14. $(4x^2 - 8x + 2) - ({}^-5x^2 + 5x + 6) =$ **$9x^2 - 13x - 4$**
15. $(10x^3 - 8x^2 + 6x - 1) - ({}^-7x^3 + 2x - 4) =$ **$17x^3 - 8x^2 + 4x + 3$**

Name________________ (6.EE.A.3, 7.EE.A.1, HSA-APR.A.1)

Multiplying Polynomials and Monomials

To multiply a multi-termed polynomial by a monomial, simply multiply each term in the polynomial by the monomial using the distributive property. Remember the properties of exponents when simplifying such problems.

Multiply.	$3x(3x^3 - 2x^2 + 8x - 7)$	
	$3x(3x^3) - 3x(2x^2) + 3x(8x) - 3x(7)$	Distribute 3x to each term.
	$9x^4 - 6x^3 + 24x^2 - 21x$	Multiply to simplify.

You may have to simplify an expression using the distributive property to multiply the polynomial by a monomial.

Simplify.	$4x^7 + 9x^4 - 7 + 2x^2(8x^5 - 5x^2)$	
	$4x^7 + 9x^4 - 7 + 2x^2(8x^5) - 2x^2(5x^2)$	Distribute the 2x.
	$4x^7 + 9x^4 - 7 + 16x^7 - 10x^4$	Multiply to begin simplifying.
	$20x^7 - x^4 - 7$	Simplify.

Multiply each expression.

1. $6x(4x - 3)$
   $24x^2 - 18x$
2. $4x(x^2 - 6x + 3)$
   $4x^3 - 24x^2 + 12x$
3. $({}^-3x^2 + 4x - 2)({}^-4x^3)$
   $12x^5 - 16x^4 + 8x^3$
4. $x({}^-5x + 2)$
   $-5x^2 + 2x$
5. ${}^-3x(5x^2 - 2x - 6)$
   $-15x^3 + 6x^2 + 18x$
6. $({}^-x^2 + 2x - 1)(5x^3)$
   $-5x^5 + 10x^4 - 5x^3$
7. ${}^-2x({}^-4x - 8)$
   $8x^2 + 16x$
8. ${}^-x^2(7x^2 - x + 4)$
   $-7x^4 + x^3 - 4x^2$
9. ${}^-5x^4(2x^2 - 8x + 6)$
   $-10x^6 + 40x^5 - 30x^4$
10. $5y(y^2 - 3y + 1)$
    $5y^3 - 15y^2 + 5y$
11. $(3x^2 - 4x)({}^-x)$
    $-3x^3 + 4x^2$
12. $7x(x^3 - x^2)$
    $7x^4 - 7x^3$

Answer Key

Name________________ (6.EE.A.3, 7.EE.A.1, HSA-APR.A.1)

Multiplying Polynomials and Monomials

$4y(y - 3) = 4y^2 - 12y$

Use the distributive property to multiply the polynomials.

1. $a(a + 8) =$ **$a^2 + 8a$**
2. $5b(4b^3 - 6b^2 - 6) =$ **$20b^4 - 30b^3 - 30b$**
3. $3x(x - 3) =$ **$3x^2 - 9x$**
4. $4a(2a + 6) =$ **$8a^2 + 24a$**
5. $y(y - 7) =$ **$y^2 - 7y$**
6. $^-2x^2(5 - 3x + 3x^2 + 4x^3) =$ **$-10x^2 + 6x^3 - 6x^4 - 8x^5$**
7. $4b(3 - b) =$ **$12b - 4b^2$**
8. $2xy(2x - 3y) =$ **$4x^2y - 6xy^2$**
9. $^-5y^2(7y - 8y^2) =$ **$-35y^3 + 40y^4$**
10. $4x^2(3x^2 - x) =$ **$12x^4 - 4x^3$**
11. $x(x^2 + x + x) =$ **$x^3 + 2x^2$**
12. $3b(4b^3 - 12b^2 - 7) =$ **$12b^4 - 36b^3 - 21b$**
13. $(^-7x^3)(3x^2 - 1) =$ **$-21x^5 + 7x^3$**
14. $^-5ab(6a - 4b) =$ **$-30a^2b + 20ab^2$**
15. $3x(x - 3) =$ **$3x^2 - 9x$**
16. $^-3x^2(4x^2 - 3x + 3) =$ **$-12x^4 + 9x^3 - 9x^2$**
17. $^-4x^2(3x^3 + 8x^2 + {}^-9x) =$ **$-12x^5 - 32x^4 + 36x^3$**
18. $(3x^4 - 5x^2 - 4)(^-3x^3) =$ **$-9x^7 + 15x^5 + 12x^3$**

© Carson-Dellosa • CD-104632 57

Name________________ (6.EE.A.3, 7.EE.A.1, HSA-APR.A.1)

Multiplying Polynomials and Monomials

Simplify each expression.

1. $^-4x^2 - 5x + 7 + 3(x^2 + 8x - 2)$ **$-x^2 + 19x + 1$**
2. $5x^2 - 3x(x - 7)$ **$2x^2 + 21x$**
3. $6x^2 + 4x + (7x - 3)2x$ **$20x^2 - 2x$**
4. $^-x^2 + 8x - 6 - 7(3x^2 - 5x + 9)$ **$-22x^2 + 43x - 69$**
5. $6x^2 + 3(x - 5) - 8x$ **$6x^2 - 5x - 15$**
6. $2x^2 + 7x + 6 - 5x(^-2x - 1)$ **$12x^2 + 12x + 6$**
7. $^-3x^2 - 3x(4x^2 - 5x + 7) - 8x^2$ **$-12x^3 + 4x^2 - 21x$**
8. $x^3 + (7x^2 - 9x - 1)x + 10x^2$ **$8x^3 + x^2 - x$**

Write and simplify an expression to find the area of each polygon. Remember that the area of a triangle can be found using $A = \frac{1}{2}bh$ and the area of a rectangle can be found using $A = lw$.

9. 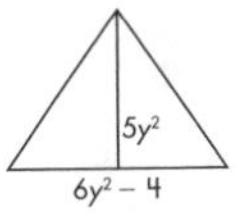

A = **$15y^4 - 10y^2$**

10. 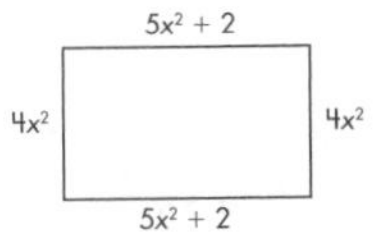

A = **$20x^4 + 8x^2$**

11. 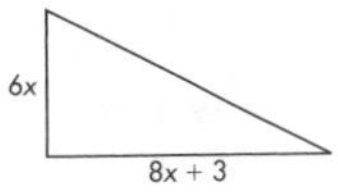

A = **$24x^2 + 9x$**

12. 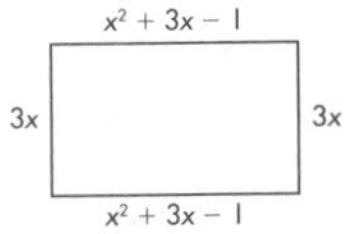

A = **$3x^3 + 9x^2 - 3x$**

58 © Carson-Dellosa • CD-104632

Name________________ (7.EE.A.1, HSA-SSE.A.2, HSA-APR.A.1, HSA-APR.C.4)

Multiplying Binomials

The easiest way to multiply binomials is to use what is called the FOIL method. This method multiplies the first terms of the binomials (F), multiplies the outer terms of the binomials (O), multiplies the inner terms of the binomials (I), multiplies the last terms of the binomials (L), and simplifies by adding like terms.

1. Multiply $(3x - 2)(4x + 3)$

$3x(4x) = 12x^2$	Multiply first terms.	F
$3x(3) = 9x$	Multiply outer terms.	O
$^-2(4x) = {}^-8x$	Multiply inner terms.	I
$^-2(3) = {}^-6$	Multiply last terms.	L
$12x^2 + 9x - 8x - 6$	Simplify by combining like terms.	
$12x^2 + x - 6$	Final product	

2. Multiply $(5x + 2)(7x - 3)$

$5x(7x) + 5x(^-3) + 2(7x) + 2(^-3)$
F O I L

$35x^2 - 15x + 14x - 6$ Multiply.

$35x^2 - x - 6$ Final product

Multiply.

1. $(x + 7)(x - 5)$ **$x^2 + 2x - 35$**
2. $(x - 5)(3x + 7)$ **$3x^2 - 8x - 35$**
3. $(9x - 1)(6x + 2)$ **$54x^2 + 12x - 2$**
4. $(7x - 8)(8x - 7)$ **$56x^2 - 113x + 56$**
5. $(x + 2)(x + 3)$ **$x^2 + 5x + 6$**
6. $(2b - 8)(3b - 7)$ **$6b^2 - 38b + 56$**
7. $(3x + 5)(^-4x - 7)$ **$-12x^2 - 41x - 35$**
8. $(^-3x - 9)(^-x - 6)$ **$3x^2 + 27x + 54$**
9. $(x - 10)(x + 1)$ **$x^2 - 9x - 10$**
10. $(2x - 3)(5x + 4)$ **$10x^2 - 7x - 12$**
11. $(^-5x + 6)(x - 2)$ **$-5x^2 + 16x - 12$**
12. $(^-4x + 4)(5x + 8)$ **$-20x^2 - 12x + 32$**
13. $(5b - 2)(3b + 2)$ **$15b^2 + 4b - 4$**
14. $(3x + y)(3x - 2y)$ **$9x^2 - 3xy - 2y^2$**
15. $(4a + 1)(3a - 3)$ **$12a^2 - 9a - 3$**
16. $(x - y)(2x + 2y)$ **$2x^2 - 2y^2$**

© Carson-Dellosa • CD-104632 59

Name________________ (7.EE.A.1, HSA-SSE.A.2, HSA-APR.A.1, HSA-APR.C.4)

Multiplying Binomials

Rules: $(a + b)(a - b) = a^2 - b^2$
$(a + b)^2 = a^2 + 2ab + b^2$
$(a - b)^2 = a^2 - 2ab + b^2$

Multiply.

1. $(2x + y)(2x - y) =$ **$4x^2 - y^2$**
2. $(b - 5)(b + 5) =$ **$b^2 - 25$**
3. $(x - y)(2x + 2y) =$ **$2x^2 - 2y^2$**
4. $(x - 4y)^2 =$ **$x^2 - 8xy + 16y^2$**
5. $(7x - 3y)(7x + 3y) =$ **$49x^2 - 9y^2$**
6. $(x + 6)(x - 6) =$ **$x^2 - 36$**
7. $(7x + y)(7x - y) =$ **$49x^2 - y^2$**
8. $(2x - 6y)^2 =$ **$4x^2 - 24xy + 36y^2$**
9. $(3a - b)(3a + b) =$ **$9a^2 - b^2$**
10. $(x + 2)(x - 2) =$ **$x^2 - 4$**
11. $(12b - 5)(12b + 5) =$ **$144b^2 - 25$**
12. $(2x - 3y)^2 =$ **$4x^2 - 12xy + 9y^2$**
13. $(7x - 5y)^2 =$ **$49x^2 - 70xy + 25y^2$**
14. $(3x + 13)(3x - 2) =$ **$9x^2 + 33x - 26$**
15. $(c + 2d)(c - 2d) =$ **$c^2 - 4d^2$**
16. $(3b + 6)(3b - 6) =$ **$9b^2 - 36$**
17. $(2b + 5a)^2 =$ **$4b^2 + 20ab + 25a^2$**
18. $(2x + 3v)^2 =$ **$4x^2 + 12xv + 9v^2$**
19. $(12 + b)(12 - b) =$ **$144 - b^2$**
20. $(5x + 7)^2 =$ **$25x^2 + 70x + 49$**

60 © Carson-Dellosa • CD-104632

Answer Key

Name ________________ (7.EE.A.1, HSA-SSE.A.2, HSA-APR.A.1, HSA-APR.C.4)

Multiplying Binomials

Multiply.

1. $(4b^2 - 4)(4b^2 + 4) =$ $\mathbf{16b^4 - 16}$
2. $(3b - 3c)^2 =$ $\mathbf{9b^2 - 18bc + 9c^2}$
3. $(x - 2y)^2 =$ $\mathbf{x^2 - 4xy + 4y^2}$
4. $(^-5x^2 + 3)(^-5x^2 - 3) =$ $\mathbf{25x^4 - 9}$
5. $(7b^2 - 3c)^2 =$ $\mathbf{49b^4 - 42b^2c + 9c^2}$
6. $(4c + 9d)^2 =$ $\mathbf{16c^2 + 72cd + 81d^2}$
7. $(5x^2 - 5y)^2 =$ $\mathbf{25x^4 - 50x^2y + 25y^2}$
8. $(x - yz)(x + yz) =$ $\mathbf{x^2 - y^2z^2}$
9. $(^-4x + 3y)^2 =$ $\mathbf{16x^2 - 24xy + 9y^2}$
10. $(4m^2 - 2n)^2 =$ $\mathbf{16m^4 - 16m^2n + 4n^2}$
11. $(x^2 - 7y)^2 =$ $\mathbf{x^2 - 14x^2y + 49y^2}$
12. $(4x^2 - 4y^2)^2 =$ $\mathbf{16x^4 - 32x^2y^2 + 16y^4}$
13. $(8x^2 - 12)(8x^2 + 12) =$ $\mathbf{64x^4 - 144}$
14. $(2b^2 - 2c^2)^2 =$ $\mathbf{4b^4 - 8b^2c^2 + 4c^4}$
15. $(4a + b)^2 =$ $\mathbf{16a^2 + 8ab + b^2}$
16. $(x^2 - 8x)(x^2 + 8x) =$ $\mathbf{x^4 - 64x^2}$
17. $(2x^2 - y^2)(2x^2 + y^2) =$ $\mathbf{4x^4 - y^4}$
18. $(3x^2 - x)(3x^2 + x) =$ $\mathbf{9x^4 - x^2}$
19. $(3a - 7b)^2 =$ $\mathbf{9a^2 - 42ab + 49b^2}$
20. $(^-6x + 3y)^2 =$ $\mathbf{36x^2 - 36xy + 9y^2}$
21. $(4x^2 - 4y^2)(4x^2 + 4y^2) =$ $\mathbf{16x^4 - 16y^4}$
22. $(^-2x^3 + 4)(2x^2 + 5) =$ $\mathbf{^-4x^5 - 10x^3 + 8x^2 + 20}$

Name ________________ (7.EE.A.1, HSA-SSE.A.2)

Factoring Polynomials

To factor a polynomial, write the polynomial as a product of other polynomials.
For example, $3x^2 - 6x$ can be written as $3x(x - 2)$.
$3x$ is the Greatest Common Factor (GCF) of $3x^2$ and $6x$.
$3x$ is a Common Monomial Factor of the terms of the binomial.
$x - 2$ is a Binomial Factor of $3x^2 - 6$.

To factor a trinomial in the form $x^2 + bx + c$, write it as the product of two binomials.
For example, $a^2 - 9a + 14$ can be written as $(a - 7)(a - 2)$.

Look for two numbers that are the product of c and whose sum is b.
If c and b are both positive, the factors will both be positive.
If c is positive and b is negative, the factors will both be negative.
If c is negative, the factors will have opposite signs.

Factor.

1. $3x^2 - 12x^3 =$ $\mathbf{3x^2(1 - 4x)}$
2. $2x^3 - x^4 =$ $\mathbf{x^3(2 - x)}$
3. $3a^5 - a^3 =$ $\mathbf{a^3(3a^2 - 1)}$
4. $x^5 + 2x^2 =$ $\mathbf{x^2(x^3 + 2)}$
5. $24b^2 + 16b =$ $\mathbf{8b(3b + 2)}$
6. $5x^3 - 7x^2 =$ $\mathbf{x^2(5x - 7)}$
7. $2x^3 + 6x^2 =$ $\mathbf{2x^2(x + 3)}$
8. $x^3 - 5x^2 =$ $\mathbf{x^2(x - 5)}$
9. $x^2 + x - 90 =$ $\mathbf{(x + 10)(x - 9)}$
10. $y^2 - 13y + 42 =$ $\mathbf{(y - 6)(y - 7)}$
11. $x^2 - x - 6 =$ $\mathbf{(x - 3)(x + 2)}$
12. $x^2 - 9x + 18 =$ $\mathbf{(x - 3)(x - 6)}$
13. $b^2 - 4b - 45 =$ $\mathbf{(b - 9)(b + 5)}$
14. $y^2 - 12y + 36 =$ $\mathbf{(y - 6)(y - 6)}$

Name ________________ (7.EE.A.1, HSA-SSE.A.2)

Factoring Polynomials

$a^2 - 9a + 14 = (a - 7)(a - 2)$
$2x^2 - 5x - 12 = (2x + 3)(x - 4)$

Factor.

1. $p^2 + 13p + 42 =$ $\mathbf{(p + 7)(p + 6)}$
2. $c^2 + c - 30 =$ $\mathbf{(c + 6)(c - 5)}$
3. $x^2 + 15xy + 44y^2 =$ $\mathbf{(x + 4y)(x + 11y)}$
4. $x^2 - 13x + 12 =$ $\mathbf{(x - 12)(x - 1)}$
5. $x^2 - 13x + 30 =$ $\mathbf{(x - 10)(x - 3)}$
6. $x^2 - 8x + 16 =$ $\mathbf{(x - 4)(x - 4)}$
7. $x^2 - 12xy + 32y^2 =$ $\mathbf{(x - 8y)(x - 4y)}$
8. $x^2 + 14x + 49 =$ $\mathbf{(x + 7)(x + 7)}$
9. $n^2 + 6n - 16 =$ $\mathbf{(n + 8)(n - 2)}$
10. $x^2 + 12x + 35 =$ $\mathbf{(x + 7)(x + 5)}$
11. $c^2 - 10c + 21 =$ $\mathbf{(c - 7)(c - 3)}$
12. $x^2 + 6x - 40 =$ $\mathbf{(x + 10)(x - 4)}$
13. $2x^2 + 9x + 10 =$ $\mathbf{(2x + 5)(x + 2)}$
14. $4x^2 - 18x + 20 =$ $\mathbf{2(2x - 5)(x - 2)}$
15. $3x^2 - 10x + 7 =$ $\mathbf{(3x - 7)(x - 1)}$
16. $3x^2 - 5x - 12 =$ $\mathbf{(3x + 4)(x - 3)}$
17. $3x^2 - 4x - 32 =$ $\mathbf{(3x + 8)(x - 4)}$
18. $5x^2 + 25x + 30 =$ $\mathbf{5(x + 3)(x + 2)}$

Name ________________ (7.EE.A.1, HSA-SSE.A.2)

Factoring Polynomials

Factor.

1. $x^2 + 4x + 4 =$ $\mathbf{(x + 2)(x + 2)}$
2. $2x^2 + 13x + 6 =$ $\mathbf{(2x + 1)(x + 6)}$
3. $x^2 + 7x + 12 =$ $\mathbf{(x + 3)(x + 4)}$
4. $6x^2 - 21x - 12 =$ $\mathbf{3(2x + 1)(x - 4)}$
5. $x^2 - 8x + 15 =$ $\mathbf{(x - 3)(x - 5)}$
6. $9x^2 - 9x - 28 =$ $\mathbf{(3x + 4)(3x - 7)}$
7. $x^2 - 15x + 56 =$ $\mathbf{(x - 7)(x - 8)}$
8. $15x^2 + 11x - 14 =$ $\mathbf{(5x + 7)(3x - 2)}$
9. $x^2 + 18x + 45 =$ $\mathbf{(x + 15)(x + 3)}$
10. $10x^2 - 28x - 6 =$ $\mathbf{2(x - 3)(5x + 1)}$
11. $x^2 + 4x - 32 =$ $\mathbf{(x + 8)(x - 4)}$
12. $7x^2 + 17x + 6 =$ $\mathbf{(7x + 3)(x + 2)}$
13. $x^2 + 2xy - 63y^2 =$ $\mathbf{(x + 9y)(x - 7y)}$
14. $2x^2 - 2x - 40 =$ $\mathbf{2(x - 5)(x + 4)}$
15. $x^2 - 14x - 72 =$ $\mathbf{(x + 4)(x - 18)}$
16. $11x^2 - 122x + 11 =$ $\mathbf{(11x - 1)(x - 11)}$
17. $x^2 - xy - 2y^2 =$ $\mathbf{(x - 2y)(x + y)}$
18. $2x^2 + 7x + 3 =$ $\mathbf{(2x + 1)(x + 3)}$
19. $x^2 + 16x + 28 =$ $\mathbf{(x + 14)(x + 2)}$
20. $12x^2 + 9x - 3 =$ $\mathbf{3(4x - 1)(x + 1)}$

Answer Key

Name ____________________ (7.EE.A.1, HSA-SSE.A.2)

Factoring Polynomials—Special Cases

Factoring Trinomials That Are Quadratic in Form
$x^4 - x^2 - 12 = (x^2)^2 - (x^2) - 12 = (x^2 - 4)(x^2 + 3)$

Factoring the Difference of Two Squares
Rule: $a^2 - b^2 = (a + b)(a - b)$ Example: $x^2 - 49 = (x + 7)(x - 7)$

Factoring Perfect Square Trinomials
Rules: $a^2 + 2ab + b^2 = (a + b)^2$ $a^2 - 2ab + b^2 = (a - b)^2$
Examples: $9x^2 + 6x + 1 = (3x + 1)^2$ $x^2 - 6x + 9 = (x - 3)^2$

Factoring the Sum or Difference of Two Cubes
Rules: $x^3 + y^3 = (x + y)(x^2 - xy + y^2)$ $x^3 - y^3 = (x - y)(x^2 + xy + y^2)$
Examples: $x^3 + 8 = (x + 2)(x^2 - 2x + 4)$ $x^3 - 8 = (x - 2)(x^2 + 2x + 4)$

Factor.

1. $2x^2 - 5x - 12 =$ $\mathbf{(2x + 3)(x - 4)}$
2. $2x^4 + 16x^2 + 30 =$ $\mathbf{(2x^2 + 10)(x^2 + 3)}$
3. $x^4 - 8x^2 + 15 =$ $\mathbf{(x^2 - 3)(x^2 - 5)}$
4. $7x^4 - 11x^2 - 6 =$ $\mathbf{(7x^2 + 3)(x^2 - 2)}$
5. $a^2 - 4 =$ $\mathbf{(a - 2)(a + 2)}$
6. $b^2 - 9 =$ $\mathbf{(b - 3)(b + 3)}$
7. $1 - 9x^2 =$ $\mathbf{(1 - 3x)(1 + 3x)}$
8. $x^2 - 25 =$ $\mathbf{(x - 5)(x + 5)}$
9. $x^2 + 14x + 49 =$ $\mathbf{(x + 7)^2}$
10. $x^2 - 2x + 1 =$ $\mathbf{(x - 1)^2}$
11. $c^2 - 6c + 9 =$ $\mathbf{(c + 3)^2}$
12. $x^2 - 4xy + 4y^2 =$ $\mathbf{(x - 2y)^2}$
13. $64x^3 + 1 =$ $\mathbf{(4x + 1)(16x^2 - 4x + 1)}$
14. $8x^3 + 27 =$ $\mathbf{(2x + 3)(4x^2 - 6x + 9)}$

Name ____________________ (7.EE.A.1, HSA-SSE.A.2)

Factoring Polynomials—Special Cases

Factor.

1. $2x^4 - 7x^2 - 15 =$ $\mathbf{(2x^2 + 3)(x^2 - 5)}$
2. $y^4 + 6y^2 - 16 =$ $\mathbf{(y^2 - 2)(y^2 + 8)}$
3. $8x^4 - 23x^2 - 3 =$ $\mathbf{(8x^2 + 1)(x^2 - 3)}$
4. $6a^6 - 5a^3b^3 - 25b^6 =$ $\mathbf{(3a^3 + 5b^3)(2a^3 - 5b^3)}$
5. $3x^4 + 20x^2 + 33 =$ $\mathbf{(3x^2 + 11)(x^2 + 3)}$
6. $4x^4y^4 - 2x^2y^2 - 56 =$ $\mathbf{(2x^2y^2 + 7)(2x^2y^2 - 8)}$
7. $y^2 - 81 =$ $\mathbf{(y - 9)(y + 9)}$
8. $c^2 - 16 =$ $\mathbf{(c - 4)(c + 4)}$
9. $a^2 - 49 =$ $\mathbf{(a - 7)(a + 7)}$
10. $49x^2 - 16y^4 =$ $\mathbf{(7x - 4y^2)(7x + 4y^2)}$
11. $16x^2 - 121 =$ $\mathbf{(4x - 11)(4x + 11)}$
12. $25 - x^2y^2 =$ $\mathbf{(5 - xy)(5 + xy)}$
13. $a^2 + 12ab + 36b^2 =$ $\mathbf{(a + 6b)^2}$
14. $49x^2 + 28x + 4 =$ $\mathbf{(7x + 2)^2}$
15. $x^2 + 14x + 49 =$ $\mathbf{(x + 7)^2}$
16. $c^2 - 20c + 100 =$ $\mathbf{(c - 10)^2}$
17. $x^3 - 1000 =$ $\mathbf{(x - 10)(x^2 + 10x + 100)}$
18. $1 - 64y^3 =$ $\mathbf{(1 - 4y)(1 + 4y + 16y^2)}$
19. $x^3 + 125 =$ $\mathbf{(x + 5)(x^2 - 5x + 25)}$
20. $x^3 - 27 =$ $\mathbf{(x - 3)(x^2 + 3x + 9)}$

Name ____________________ (7.EE.A.1, HSA-SSE.A.2)

Factoring Polynomials—Special Cases

Factor.

1. $y^4 - y^2 - 12 =$ $\mathbf{(y^2 + 3)(y^2 - 4)}$
2. $x^4y^4 - 19x^2y^2 + 34 =$ $\mathbf{(x^2y^2 - 2)(x^2y^2 - 17)}$
3. $2x^4y^4 - 17x^2y^2 - 30 =$ $\mathbf{(2x^2y^2 + 3)(x^2y^2 - 10)}$
4. $x^4y^4 - 8x^2y^2 + 12 =$ $\mathbf{(x^2y^2 - 6)(x^2y^2 - 2)}$
5. $64 - x^4y^4 =$ $\mathbf{(8 - x^2y^2)(8 + x^2y^2)}$
6. $y^2 - 64 =$ $\mathbf{(y - 8)(y + 8)}$
7. $81x^2 - 4 =$ $\mathbf{(9x - 2)(9x + 2)}$
8. $16 - 81x^2 =$ $\mathbf{(4 - 9x)(4 + 9x)}$
9. $x^2y^2 - 121 =$ $\mathbf{(xy - 11)(xy + 11)}$
10. $49x^2 - 36 =$ $\mathbf{(7x - 6)(7x + 6)}$
11. $y^2 - 22y + 121 =$ $\mathbf{(y - 11)^2}$
12. $25a^2 - 40ab + 16b^2 =$ $\mathbf{(5a - 4b)^2}$
13. $9x^2 - 12x + 4 =$ $\mathbf{(3x - 2)^2}$
14. $16x^2 - 40x + 25 =$ $\mathbf{(4x - 5)^2}$
15. $9x^2 + 12x + 4 =$ $\mathbf{(3x + 2)^2}$
16. $x^2 - 14x + 49 =$ $\mathbf{(x - 7)^2}$
17. $27x^3 - 64 =$ $\mathbf{(3x - 4)(9x^2 + 12x + 16)}$
18. $27x^3 - 27 =$ $\mathbf{27(x - 1)(x^2 + x + 1)}$
19. $125a^3 - 8b^3 =$ $\mathbf{(5a - 2b)(25a^2 + 10ab + 4b^2)}$
20. $64x^3 - y^3 =$ $\mathbf{(4x - y)(16x^2 + 4xy + y^2)}$
21. $125x^3 - 64y^3 =$ $\mathbf{(5x - 4y)(25x^2 + 20xy + 16y^2)}$
22. $64x^3 + 27 =$ $\mathbf{(4x + 3)(16x^2 - 12x + 9)}$

Name ____________________ (HSA-SSE.B.3a, HSA-REI.B.4b)

Solving Equations by Factoring

The **Multiplication Property of Zero**: The product of a number and zero is zero.

The **Principle of Zero Products**: If the product of two factors is zero, then at least one of the factors must be zero. This principle is used in solving equations.

Solve: $(x - 4)(x - 5) = 0$
$x - 4 = 0$ $x - 5 = 0$
$x = 4$ $x = 5$
The solutions are 4 and 5.

If $(x - 4)(x - 5) = 0$, then $(x - 4) = 0$ or $(x - 5) = 0$.
$x = 4$: $(4 - 4)(4 - 5) = 0$; $(0)(-1) = 0$; $0 = 0$
$x = 5$: $(5 - 4)(5 - 5) = 0$; $(1)(0) = 0$; $0 = 0$

Write the solutions for each variable.

1. $x(x + 6) = 0$ **0, −6**
2. $b^2 - 81 = 0$ **−9, 9**
3. $(27 - y)(y - 2) = 0$ **2, 27**
4. $z^2 - 1 = 0$ **−1, 1**
5. $(y - 4)(y - 8) = 0$ **4, 8**
6. $y(y - 11) = 0$ **0, 11**
7. $8b^2 - 32 = 0$ **2, −2**
8. $x^2 - x - 6 = 0$ **−2, 3**
9. $x^2 - 4x - 21 = 0$ **−3, 7**
10. $m^2 - 144 = 0$ **−12, 12**
11. $(y + 5)(y + 6) = 0$ **−5, −6**
12. $(2x + 4)(x + 7) = 0$ **−7, −2**
13. $z^2 - 9 = 0$ **−3, 3**
14. $10x^2 - 10x = 0$ **1, 0**
15. $2x^2 - 6x = x - 3$ $\mathbf{\frac{1}{2}, 3}$
16. $4y(3y - 2) = 0$ $\mathbf{0, \frac{2}{3}}$
17. $(4y - 1)(y + 2) = 0$ $\mathbf{\frac{1}{4}, -2}$
18. $x^2 - 5x + 6 = 0$ **2, 3**

Answer Key

Name____________________ (HSA-SSE.B.3a, HSA-REI.B.4b)

Solving Equations by Factoring

$x^2 - 8x = {}^-16$
$x^2 - 8x + 16 = 0$
$(x - 4)(x - 4) = 0$
$x - 4 = 0$
$x = 4$
The solution is 4.

Solve by factoring.

1. $y^2 + 9y = 0$ **−9, 0**
2. $x - 16 = x(x - 7)$ **4**
3. $x - 6 = x(x - 4)$ **2, 3**
4. $x^2 + 7x = 0$ **0, −7**
5. $x^2 - 4x = 0$ **0, 4**
6. $x + 8 = x(x + 3)$ **−4, 2**
7. $y^2 - y - 6 = 0$ **−2, 3**
8. $a^2 - 36 = 0$ **−6, 6**
9. $y^2 + 15 = 8y$ **3, 5**
10. $a^2 - 7a = {}^-12$ **3, 4**
11. $y^2 + 36y = 0$ **0, −36**
12. $3u^2 - 12u - 15 = 0$ **−1, 5**
13. $y^2 - 8y + 12 = 0$ **2, 6**
14. $5a^2 + 25a = 0$ **−5, 0**
15. $6x^2 + 18x = 0$ **−3, 0**
16. $2x^2 + x = 6$ **−2, $\frac{3}{2}$**
17. $x^2 - 5x - 6 = 0$ **−1, 6**
18. $4x^2 + 16x = 0$ **0, −4**
19. $3x^2 - 9x = 0$ **0, 3**
20. $y^2 + 5y - 6 = 0$ **−6, 1**

Name____________________ (HSA-SSE.B.3a, HSA-REI.B.4b)

Solving Equations by Factoring

Solve by factoring.

1. $a^2 - 8a = 0$ **0, 8**
2. $x^2 = 3x + 4$ **−1, 4**
3. $4a^2 + 15a - 4 = 0$ **$\frac{1}{4}$, −4**
4. $x^2 - x - 6 = 0$ **−2, 3**
5. $3x^2 - 13x + 4 = 0$ **$\frac{1}{3}$, 4**
6. $6x^2 = 23x + 18$ **$-\frac{2}{3}$, $\frac{9}{2}$**
7. $x^2 + 7x + 12 = 0$ **−4, −3**
8. $x^2 + 5x - 6 = 0$ **−6, 1**
9. $x^2 = 6x + 7$ **−1, 7**
10. $x^2 = 10x - 25$ **5**
11. $x^2 + 3x - 10 = 0$ **−5, 2**
12. $x^2 - 6x + 9 = 0$ **3**
13. $y^2 - 3y + 2 = 0$ **1, 2**
14. $2x^2 - 9x + 9 = 0$ **$\frac{3}{2}$, 3**
15. $r^2 - 15r = 16$ **−1, 16**
16. $x^2 + 7x + 10 = 0$ **−5, −2**
17. $3x^2 - 2x - 8 = 0$ **$-\frac{4}{3}$, 2**
18. $2a^2 + 4a - 6 = 0$ **−3, 1**
19. $x^2 + 3x - 4 = 0$ **−4, 1**
20. $4a^2 + 9a + 2 = 0$ **−2, $-\frac{1}{4}$**
21. $9x^2 = 18x + 0$ **0, 2**
22. $2x^2 = 9x + 5$ **$-\frac{1}{2}$, 5**

Name____________________ (7.EE.B.4, HSA-SSE.B.3a, HSA-REI.B.4b)

Writing and Solving Quadratic Equations

Problem solving often involves quadratic equations.

The square of a number is 8 less than 6 times the number. Find the numbers that make the sentence true.

$x^2 = 6x - 8$	First, set up the equation.
$x^2 - 6x + 8 = 0$	Put the equation in standard form.
$(x - 2)(x - 4) = 0$	Factor.
$x - 2 = 0$ or $x - 4 = 0$	Solve for x.
$x = 2$ or $x = 4$	So, the numbers are 2 and 4.

The product of a number and 10 more than 4 times the number is 50. Find the numbers that make the sentence true.

$x(4x + 10) = 50$	First, set up the equation.
$4x^2 + 10x - 50 = 0$	Put the equation in standard form.
$2(2x - 5)(x + 5) = 0$	Factor.
Since $2 \neq 0$, then $2x - 5 = 0$ or $x + 5 = 0$	Solve for x.
$x = \frac{5}{2}$ and $x = {}^-5$	So, the numbers are $\frac{5}{2}$ and $^-5$.

Write an equation for each word problem and solve it.

1. Six less than 3 times a number is 12 less than twice the number.
   Equation **$3x - 6 = 2x - 12$** Solution **−6**
2. The square of a number is 8 more than twice the number.
   Equation **$x^2 = 2x + 8$** Solution **4, −2**
3. The sum of the square of a number and 4 times the number is 12.
   Equation **$x^2 + 4x = 12$** Solution **−6, 2**
4. Twice the square of a number is 4 more than twice the number.
   Equation **$2x^2 = 2x + 4$** Solution **−1, 2**
5. Three times the square of a number is 5 less than 16 times the number.
   Equation **$3x^2 = 16x - 5$** Solution **$\frac{1}{3}$, 5**

Name____________________ (7.EE.B.4, HSA-SSE.B.3a, HSA-REI.B.4b)

Writing and Solving Quadratic Equations

The length of a rectangle is 3 inches longer than the width. The area of the rectangle is 40 square inches. Find the length and width of the rectangle.

Width of rectangle: w
Length of rectangle: $w + 3$
$A = lw$
$40 = (w + 3)(w)$
$40 = w^2 + 3w$
$0 = w^2 + 3w - 40$
$0 = (w + 8)(w - 5)$
$w + 8 = 0$ or $w - 5 = 0$
$w = {}^-8 \quad w = 5$

Since the width cannot be a negative number, the width is 5.

$l = 5 + 3 = 8$

The length is 8 inches, and the width is 5 inches.

Write an equation for each word problem and solve it.

1. The area of a square is 121 m². Find the length of the sides of the square.
   Equation **$s^2 = 121$** Solution **$s = 11$ m**
2. The area of a rectangle is 72 m². Its length is twice its width. Find the length and width of the rectangle.
   Equation **$2w(w) = 72$** Solution **$l = 12$ m, $w = 6$ m**
3. The area of a rectangle is 36 cm². Its width is 4 times its length. Find the length and width of the rectangle.
   Equation **$(l)4l = 36$** Solution **$l = 3$ cm, $w = 12$ cm**
4. The width of a rectangle is 5 more than twice its length. The area of the rectangle is 33 in.². Find the dimensions of the rectangle.
   Equation **$(2l + 5)(l) = 33$** Solution **$l = 3$ in, $w = 11$ in.**
5. The length of a rectangle is 4 more than twice its width. The area of the rectangle is 96 ft.². Find its dimensions.
   Equation **$w(2w + 4) = 96$** Solution **$l = 16$ ft, $w = 6$ ft.**

Answer Key

Name ______ (7.EE.B.4, HSA-SSE.B.3a, HSA-REI.B.4b)

Writing and Solving Quadratic Equations

Write an equation for each word problem and solve it.

1. The length of a rectangle is 5 inches longer than the width. The area of the rectangle is 50 square inches. Find the length and width.
   Equation **$w(w + 5) = 50$** Solution **$l = 10$ in, $w = 5$ in.**

2. The length of a rectangle is 2 more than 8 times its width. Its area is 36 cm². Find the width and length.
   Equation **$w(8w + 2) = 36$** Solution **$l = 18$ cm, $w = 2$ cm**

3. The sum of a number and its square is 42. Find the numbers.
   Equation **$x + x^2 = 42$** Solution **$x = -7, 6$**

4. The sum of a number and its square is 56. Find the numbers.
   Equation **$x + x^2 = 56$** Solution **$x = -8, 7$**

5. The length of a rectangle is twice its width. Its area is 32 m². Find the length and width.
   Equation **$w(2w) = 32$** Solution **$l = 8$ m, $w = 4$ m**

6. The square of a number is 80 more than 2 times the number. Find the numbers.
   Equation **$x^2 = 2x + 80$** Solution **$x = 10, -8$**

7. The square of a number is 48 more than 2 times the number. Find the numbers.
   Equation **$x^2 = 2x + 48$** Solution **$x = 8, -6$**

8. The width of a rectangle is 2/3 the length. The area is 54 in². Find the length and width.
   Equation **$\frac{2}{3}l^2 = 54$** Solution **$l = 9$ in, $w = 6$ in.**

© Carson-Dellosa • CD-104632 73

Name ______ (HSA-APR.D.6)

Dividing Polynomials

Simplify using long division: $(x^2 + 6x + 5) \div (x + 1)$

$$\frac{(x^2 + 6x + 5)}{(x + 1)}$$

$$\begin{array}{r} x + 5 \\ x + 1\overline{)x^2 + 6x + 5} \\ \underline{-x^2 + 1x} \\ 5x + 5 \\ \underline{-5x + 5} \\ 0 \end{array}$$

Divide by using long division.

1. $(x^2 + 5x + 6) \div (x + 3) =$ **$x + 2$**
2. $(x^2 + 4x - 21) \div (x - 3) =$ **$x + 7$**
3. $(x^2 - 3x - 40) \div (x + 5) =$ **$x - 8$**
4. $(x^2 - x - 42) \div (x + 6) =$ **$x - 7$**
5. $(x^2 - 8x + 16) \div (x - 4) =$ **$x - 4$**
6. $(x^2 + 2x - 35) \div (x + 7) =$ **$x - 5$**
7. $(x^2 - 6x + 9) \div (x - 3) =$ **$x - 3$**
8. $(x^2 + 5x + 4) \div (x + 1) =$ **$x + 4$**
9. $(x^2 + 7x + 10) \div (x + 2) =$ **$x + 5$**
10. $(x^2 + 9x + 8) \div (x + 8) =$ **$x + 1$**

74 © Carson-Dellosa • CD-104632

Name ______ (HSA-APR.D.6)

Dividing Polynomials

Simplify using synthetic division: $(2x^3 + 3x^2 - 4x + 8) \div (x + 3)$

$$\begin{array}{r|rrrr} -3 & 2 & 3 & -4 & 8 \\ & & -6 & 9 & -15 \\ \hline & 2 & -3 & 5 & -7 \end{array}$$

$= 2x^2 - 3x + 5 - \frac{7}{x + 3}$

Divide by using synthetic division.

1. $(x^2 + x - 2) \div (x + 2) =$ **$x - 1$**
2. $(5x^2 - 12x - 9) \div (x - 3) =$ **$5x + 3$**
3. $(3x^2 - 5) \div (x - 1) =$ **$3x + 3 - \frac{2}{(x - 1)}$**
4. $(3x^3 + 8x^2 + 9x + 10) \div (x + 2) =$ **$3x^2 + 2x + 5$**
5. $(x^3 - 4x^2 - 36x - 16) \div (x + 4) =$ **$x^2 - 8x - 4$**
6. $(3x^2 - 7x + 6) \div (x - 3) =$ **$3x + 2 + \frac{12}{(x - 3)}$**
7. $(4x^2 + 9x + 6) \div (x + 1) =$ **$4x + 5 + \frac{1}{(x + 1)}$**
8. $(4x^2 + 23x + 28) \div (x + 4) =$ **$4x + 7$**
9. $(3x^2 + 19x + 20) \div (x + 5) =$ **$3x + 4$**
10. $(x^2 + 14x + 45) \div (x + 5) =$ **$x + 9$**

© Carson-Dellosa • CD-104632 75

Name ______ (HSA-APR.D.6)

Dividing Polynomials

Divide.

1. $\frac{x^2 + 9x + 8}{x - 1} =$ **$x - 8$**
2. $\frac{2x^2 - 5x - 3}{x - 3} =$ **$2x + 1$**
3. $\frac{4x^2 - 7x - 2}{4x + 1} =$ **$x - 2$**
4. $\frac{2x^2 + x - 3}{x - 1} =$ **$2x + 3$**
5. $\frac{21x^2 + 22x - 8}{3x + 4} =$ **$7x - 2$**
6. $\frac{3x^2 - 27}{x + 3} =$ **$3x - 9$**
7. $\frac{9x^2 - 27x - 36}{x - 4} =$ **$9x + 9$**
8. $\frac{5x^2 + 43x - 18}{x + 9} =$ **$5x - 2$**
9. $\frac{x^2 - 4x - 45}{x + 5} =$ **$x - 9$**
10. $\frac{2x^2 - x - 21}{x + 3} =$ **$2x - 7$**
11. $(3x^3 - 13x^2 - 13x + 15) \div (x - 5) =$ **$3x^2 + 2x - 3$**
12. $(2x^3 - 12x^2 + 5x - 27) \div (x - 6) =$ **$2x^2 + 5 + \frac{3}{(x - 6)}$**
13. $(3x^2 - 75) \div (x - 5) =$ **$3x + 15$**
14. $(2x^2 + 7x - 10) \div (x + 1) =$ **$2x + 5 - \frac{15}{(x + 1)}$**

76 © Carson-Dellosa • CD-104632

Answer Key

Name ____________________ HSA-APR.D.7

Operations with Rational Expressions

Add or subtract:

$$\frac{7x-12}{2x^2+5x-12} - \frac{3x-6}{2x^2+5x-12} = \frac{(7x-12)-(3x-6)}{2x^2+5x-12} = \frac{4x-6}{(2x-3)(x+4)} = \frac{2(2x-3)}{(2x-3)(x+4)} = \frac{2}{x+4}$$

Multiply:

$$\frac{4x+16}{12x+48} \cdot \frac{(x-4)}{2x} = \frac{4(x+4)(x-4)}{24x(x+4)} = \frac{x-4}{6x}$$

Divide:

$$\frac{x+2}{x+3} \div \frac{x^2-4}{x-2} = \frac{x+2}{x+3} \cdot \frac{x-2}{(x+2)(x-2)} = \frac{1}{x+3}$$

Simplify.

1. $\frac{x^2+x-6}{x+1} \cdot \frac{x+1}{x^2-9} =$ $\mathbf{\frac{(x-2)}{(x-3)}}$
2. $\frac{x^2-1}{x^2-2x-3} \cdot \frac{x+4}{6x-6} =$ $\mathbf{\frac{(x+4)}{6(x-3)}}$
3. $\frac{x-7}{x+2} \div \frac{x^2-49}{x^2+9x+14} =$ **1**
4. $\frac{x+2}{4x(x-6)} \div \frac{x^2-4}{8x(x-6)} =$ $\mathbf{\frac{2}{(x-2)}}$
5. $\frac{2}{x+2} + \frac{6x}{x^2-4} =$ $\mathbf{\frac{8x-4}{[(x-2)(x+2)]}}$
6. $\frac{2}{4xy} + \frac{14}{3xy} - \frac{9}{2xy} =$ $\mathbf{\frac{11}{12xy}}$

© Carson-Dellosa • CD-104632 77

Name ____________________ HSA-APR.D.7

Operations with Rational Expressions

Simplify.

1. $\frac{x+2}{x-3} \cdot \frac{x^2-8x+15}{5x-25} =$ $\mathbf{\frac{(x+2)}{5}}$
2. $\frac{12x^2y^4}{36ab^3} \cdot \frac{6a^2v^3}{48xy^4} =$ $\mathbf{\frac{xav^3}{24b^3}}$
3. $\frac{x^2+6x+8}{x^2-16} \cdot \frac{3x-12}{4x+4} =$ $\mathbf{\frac{3(x+2)}{4(x+1)}}$
4. $\frac{x^2-2x-8}{x^2-4} \cdot \frac{x-2}{x+3} =$ $\mathbf{\frac{(x-4)}{(x+3)}}$
5. $\frac{x^2+3x+2}{x+7} \cdot \frac{x^2+9x+14}{x^2+4x+4} =$ $\mathbf{x+1}$
6. $\frac{2x^2+12x+18}{x^2+5x-6} \div \frac{2x+6}{x-1} =$ $\mathbf{\frac{(x+3)}{(x+6)}}$
7. $\frac{2x^2+6x}{x^2+2x} \div \frac{x^2-9}{4x-12} =$ $\mathbf{\frac{8}{(x+2)}}$
8. $\frac{15x^4y^2}{5xy} \div \frac{10x^3y}{5y^2} =$ $\mathbf{\frac{3y^2}{2}}$
9. $\frac{x^2+8x}{x^2+14x+48} \div \frac{x^2+x}{x^2} =$ $\mathbf{\frac{x^2}{(x+6)(x+1)}}$
10. $\frac{3x^2+6x}{x^2+6x} \div \frac{x^2-4}{2x-4} =$ $\mathbf{\frac{6}{(x+6)}}$
11. $\frac{x}{x-4} + \frac{4}{x^2-x-12} =$ $\mathbf{\frac{x^2+3x+4}{[(x-4)(x+3)]}}$
12. $\frac{4}{3x-8} - \frac{x}{4x-7} =$ $\mathbf{\frac{-3x^2+24x-28}{[(4x-7)(3x-8)]}}$
13. $-\frac{4}{2x^2} + \frac{5}{2x^2} + \frac{8}{3x^2} =$ $\mathbf{\frac{19}{6x^2}}$
14. $\frac{4x}{x^2+x-2} - \frac{4x}{x^2+x-2} =$ **0**

78 © Carson-Dellosa • CD-104632

Name ____________________ HSA-APR.D.7

Operations with Rational Expressions

Simplify.

1. $\frac{2x^2-32}{2x+8} \cdot \frac{x^2-9}{x^2-3x-4} =$ $\mathbf{\frac{(x^2-9)}{(x+1)}}$
2. $\frac{14x^2y^4}{42a^2b^4} \cdot \frac{28a^2b^3}{35x^3y^4} =$ $\mathbf{\frac{4}{15bx}}$
3. $\frac{x^2-100}{x-5} \cdot \frac{x+5}{x^2-5x-50} =$ $\mathbf{\frac{(x+10)}{(x-5)}}$
4. $\frac{x^2-12x+35}{x^2-5x-14} \cdot \frac{x^2+7x+10}{x^2-25} =$ **1**
5. $\frac{9x^2-25}{4x^2+4x-3} \cdot \frac{40x^2-10x}{3x+5} =$ $\mathbf{\frac{10x(12x^2-24x+5)}{(4x^2+4x-3)}}$
6. $\frac{x+4}{6x^2+24} \cdot \frac{2x^3-8x}{x^2+4x} =$ $\mathbf{\frac{1}{3}}$
7. $\frac{x^2-7x}{x^2-14x+49} \div \frac{2x^2+6x}{x^2+x-56} =$ $\mathbf{\frac{(x+8)}{2(x+3)}}$
8. $\frac{x^2-16}{x^2+7x+12} \div \frac{5x-20}{x+3} =$ $\mathbf{\frac{1}{5}}$
9. $\frac{27x^6y^2}{9x^3y} \div \frac{4x^5y^3}{16x^4y^2} =$ $\mathbf{12x^2}$
10. $\frac{x^2+9x-10}{x^2+5x-14} \div \frac{3x+30}{2x-4} =$ $\mathbf{\frac{(2x-2)}{(3x+21)}}$
11. $\frac{24a^4b^2}{8a^2b} \div \frac{12ab^3}{16a^3b} =$ $\mathbf{\frac{4a^2}{b}}$
12. $\frac{11x}{x^2-6x-7} + \frac{5x}{x^2+9x+8} =$ $\mathbf{\frac{(16x^2+53x)}{[(x-7)(x+1)(x+8)]}}$
13. $\frac{3x}{x^2-4} + \frac{5x}{x^2-4} =$ $\mathbf{\frac{8x}{(x+2)(x-2)}}$
14. $-\frac{10}{4x^2} + \frac{6}{4x^2} + \frac{8}{4x^2} =$ $\mathbf{\frac{1}{x^2}}$
15. $\frac{5}{2xy} + \frac{5}{4xy} - \frac{12}{6xy} =$ $\mathbf{\frac{7}{4xy}}$
16. $\frac{8}{4x+16} - \frac{1}{x+4} - \frac{3}{4} =$ $\mathbf{\frac{-3x-8}{4(x+4)}}$

© Carson-Dellosa • CD-104632 79

Name ____________________ 7.RP.A.1

Ratios and Proportions

Solve the following ratio for x.

$\frac{x}{5} = \frac{4}{10}$ → Take cross products and solve. → $5 \cdot 4 = 20$, $x \cdot 10 = 10x$ → $10x = 20$ → $\frac{10x}{10} = \frac{20}{10}$ → $x = 2$

Solve.

1. $\frac{4}{(x-3)} = \frac{28}{49}$ $\mathbf{x = 10}$
2. $\frac{(5+x)}{10} = \frac{2}{5}$ $\mathbf{x = -1}$
3. $\frac{x}{30} = \frac{7}{10}$ $\mathbf{x = 21}$
4. $\frac{(x-2)}{16} = \frac{x}{4}$ $\mathbf{x = -\frac{2}{3}}$
5. $\frac{2}{x} = \frac{6}{30}$ $\mathbf{x = 10}$
6. $\frac{(x+1)}{7} = \frac{6}{14}$ $\mathbf{x = 2}$
7. $\frac{x}{15} = \frac{5}{75}$ $\mathbf{x = 1}$
8. $\frac{x}{20} = \frac{2}{10}$ $\mathbf{x = 4}$
9. $\frac{x}{6} = \frac{(x-3)}{12}$ $\mathbf{x = -3}$
10. $\frac{x}{5} = \frac{12}{6}$ $\mathbf{x = 10}$
11. $\frac{6}{(x+5)} = \frac{18}{24}$ $\mathbf{x = 3}$
12. $\frac{5}{15} = \frac{x}{9}$ $\mathbf{x = 3}$
13. $\frac{+10}{} = \frac{5}{2}$ $\mathbf{x = \frac{25}{2}}$
14. $\frac{x}{3} = \frac{12}{27}$ $\mathbf{x = \frac{4}{3}}$

80 © Carson-Dellosa • CD-104632

Answer Key

Name ____________________ (6.RP.A.3, 7.RP.A.1, 7.RP.A.2c, 7.RP.A.3)

Ratios and Proportions

Three liters of soda cost $3.00. At this rate, how much would 10 liters of soda cost? To find the cost, write and solve a ratio using *x* to represent the cost.

$\frac{\text{liters}}{\text{cost}} \longrightarrow \frac{3}{3.00} = \frac{10}{x} \longrightarrow 3x = 10(3.00) \longrightarrow 3x = 30 \longrightarrow \frac{3x}{3} = \frac{30}{3} \longrightarrow x = 10 \longrightarrow$ The cost of 10 liters of soda is $10.00.

Solve.

1. $\frac{4}{x+8} = \frac{3}{x}$ **x = 24**
2. $\frac{x}{4} = \frac{21}{7}$ **x = 12**
3. $\frac{9}{3} = \frac{6}{x}$ **x = 2**
4. $\frac{x-3}{4} = \frac{x-4}{3}$ **x = 7**
5. $\frac{x}{4} = \frac{8}{2x}$ **x = 4**
6. $\frac{12}{x} = 3$ **x = 4**

Set up a proportion and solve.

7. A copy machine can print 120 pages per minute. At this rate, how many minutes will it take to make 840 copies?
   $\frac{\textbf{minutes}}{\textbf{copies}} = \frac{1}{120} = \frac{x}{840}$ **x = 7 minutes**
8. Two gallons of fruit juice will serve 35 people. How much fruit juice is needed to serve 105 people?
   $\frac{\textbf{gallons}}{\textbf{people}} = \frac{2}{35} = \frac{x}{105}$ **x = 6 gallons**
9. A stock investment of $4,000 earns $360 in interest each year. At the same rate, how much interest will a person earn if he invests $6,000?
   $\frac{\textbf{interest}}{\textbf{investment}} = \frac{\$360}{\$4,000} = \frac{x}{\$6,000}$ **x = $540**

Name ____________________ (6.RP.A.3, 7.RP.A.1, 7.RP.A.2c, 7.RP.A.3)

Ratios and Proportions

Set up a proportion and solve.

1. An investment of $36,000 earns $900 each year. At the same rate, how much money must be invested to earn $1,200 each year?
   $\frac{\textbf{investment}}{\textbf{earning}} = \frac{\$36,000}{\$900} = \frac{x}{\$1,200}$ **x = $48,000**
2. The sales tax on a $15,000 car is $540. At this rate, what is the tax on a $32,000 car?
   $\frac{\textbf{tax}}{\textbf{value}} = \frac{\$540}{\$15,000} = \frac{x}{\$32,000}$ **x = $1,152**
3. Six gallons of paint will cover 120 doors. At this rate, how many gallons of paint are needed to cover 480 doors?
   $\frac{\textbf{gallons}}{\textbf{doors}} = \frac{6}{120} = \frac{x}{480}$ **x = 24 gallons**
4. A lawnmower can cut 1 acre on 0.5 gallons of gasoline. At this rate, how much gasoline is needed to cut 3.5 acres?
   $\frac{\textbf{gallons}}{\textbf{acres}} = \frac{0.5}{1} = \frac{x}{3.5}$ **x = 1.75 gallons**
5. An aerobics instructor burns 400 calories in 1 hour. How many hours would the instructor have to do aerobics to burn 660 calories?
   $\frac{\textbf{hours}}{\textbf{calories}} = \frac{1}{400} = \frac{x}{660}$ **x = 1.65 hours**
6. The real estate tax for a house that costs $56,000 is $1,400. At this rate, what is the value of a house for which the real estate tax is $1,800?
   $\frac{\textbf{value}}{\textbf{tax}} = \frac{\$56,000}{\$1,400} = \frac{x}{\$1,800}$ **x = $72,000**
7. One hundred thirty-six tiles are required to tile a 36 ft.² area. At this rate, how many tiles are required to tile a 288 ft.² area?
   $\frac{\textbf{tiles}}{\textbf{area}} = \frac{136}{36} = \frac{x}{288}$ **x = 1,088 tiles**

Name ____________________ (5.G.A.1, 6.NS.C.6b, 6.NS.C.6c)

Graphing Ordered Pairs

Start at the origin (0, 0).
(x, y) = (1, ⁻2) right 1 and down 2
(x, y) = (⁻3, 4) left 3 and up 4
(x, y) = (2, ⁻2) right 2 and down 2

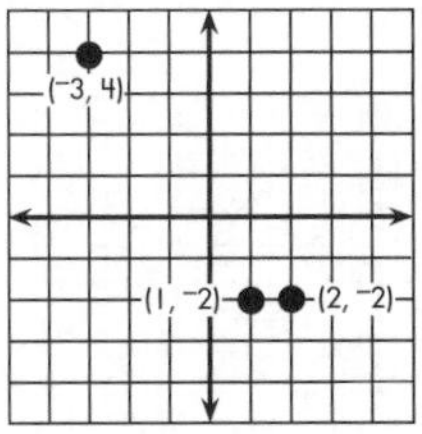

Label the following points. Start at the origin (0, 0).

A (6, 2)
B (6, 3)
C (5, 1)
D (5, 0)
E (1, ⁻3)
F (2, ⁻6)
G (5, 4)
H (2, 4)
I (4, 3)
J (1, ⁻2)
K (5, ⁻3)
L (1, 4)
M (3, 1)

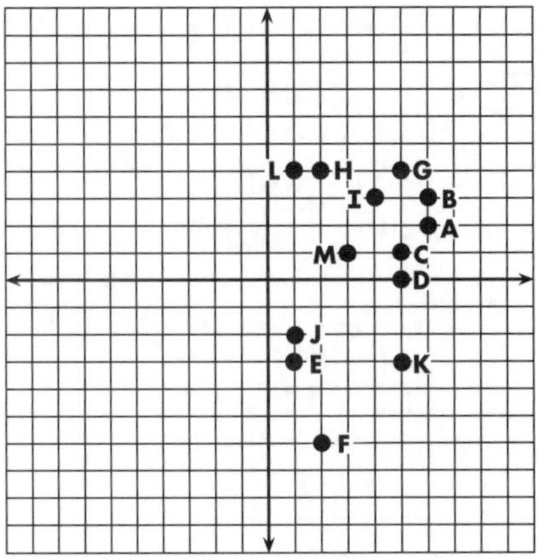

Name ____________________ (5.G.A.1, 6.NS.C.6b, 6.NS.C.6c)

Graphing Ordered Pairs

Graph and label the following points.

A (⁻1, 2) B (⁻1, 0) C (⁻1, ⁻6) D (⁻4, 2)
E (⁻6, 2) F (⁻5, 6) G (⁻4, 1) H (2, ⁻8)
I (6, 6) J (⁻7, 5) K (⁻1, ⁻1) L (⁻3, ⁻3)
M (6, ⁻4)

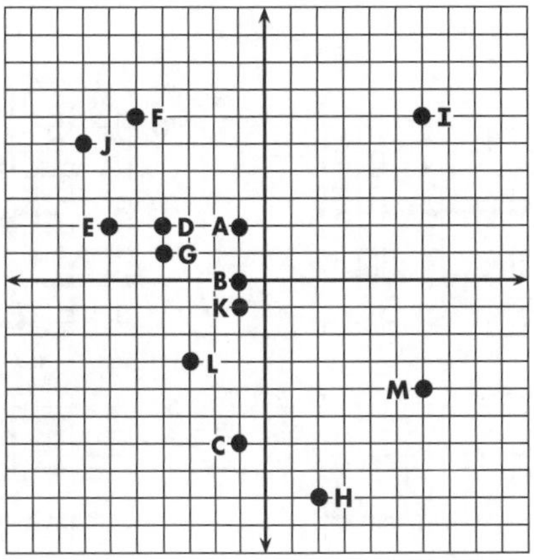

Answer Key

Name______________________ (6.NS.C.6, 6.EE.B.6, 8.F.A.1, HSA-REI.D.10)

Graphing Ordered Pairs

Solve for each x value. Graph each ordered pair.

1. $x + y = 4$

x	y
4	**0**
5	**−1**
6	**−2**

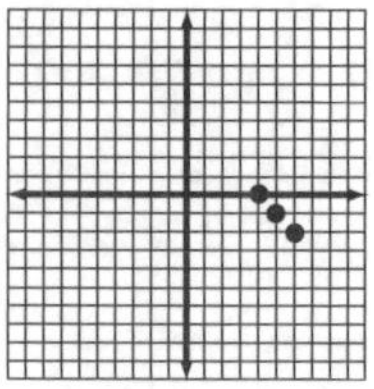

2. $y = 3 - x$

x	y
5	**−2**
−2	**5**
0	**3**

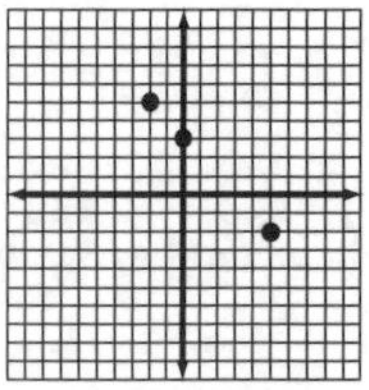

3. $4x + y = 6$

x	y
0	**6**
1	**2**
2	**−2**

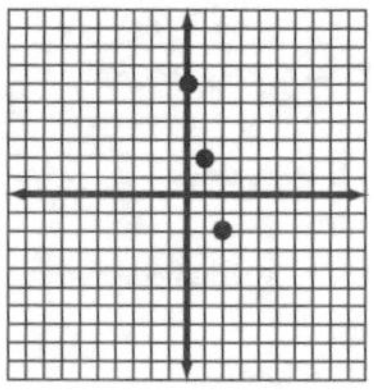

4. $y = 2x - 7$

x	y
3	**−1**
4	**1**
5	**3**

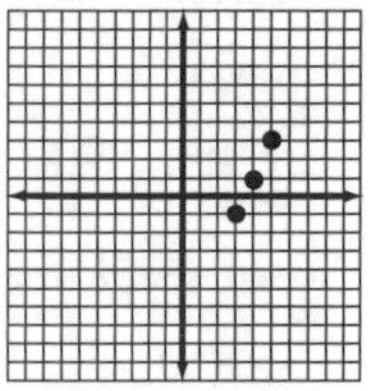

5. $y = x + 3$

x	y
2	**5**
0	**3**
−3	**0**

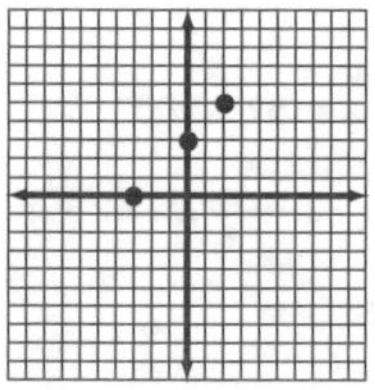

6. $y = 9 - x$

x	y
0	**9**
4	**5**
3	**6**

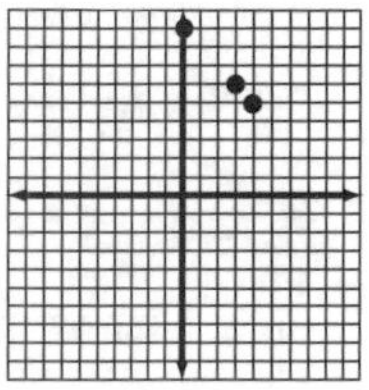

Name______________________ (8.EE.B.6, 8.F.A.1)

Graphing Linear Equations

One way to graph a linear equation is to find the x- and y-intercepts. To find the x-intercept, let $y = 0$. To find the y-intercept, let $x = 0$. Plot the two points and connect with a straight line.

Graph $2x - y = 2$ by using the x- and y-intercepts.

x-intercept	**y-intercept**
$2x - y = 2$	$2x - y = 2$
$2x - 0 = 2$	$2(0) - y = 2$
$2x = 2$	$^{-}y = 2$
$x = 1$	$y = {}^{-}2$
(1, 0)	(0, −2)

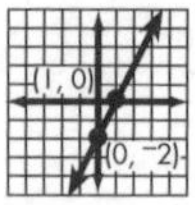

Find the x- and y-intercepts and graph.

1. $3x - 9y = 18$
(0, −2)
(6, 0)

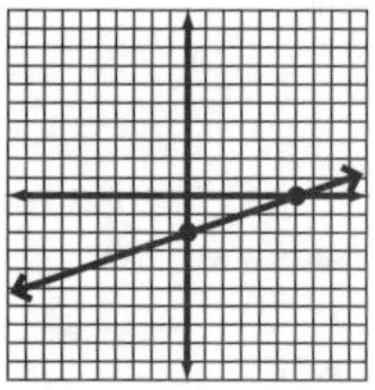

2. $4x + 2y = {}^{-}8$
(0, −4)
(−2, 0)

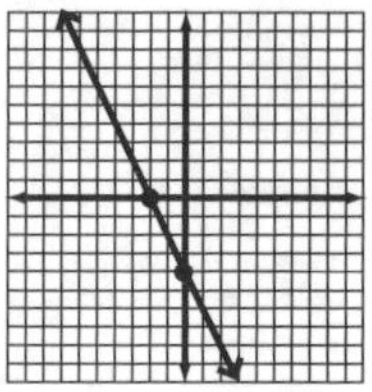

3. $2x + 4y = 20$
(0, 5)
(10, 0)

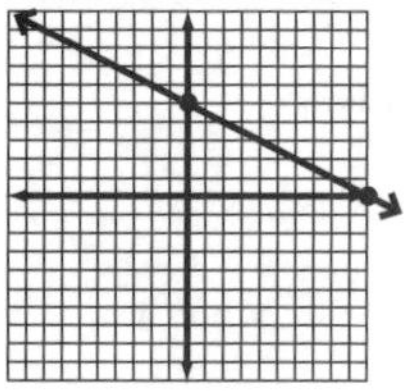

4. $x + 5y = 10$
(0, 2)
(10, 0)

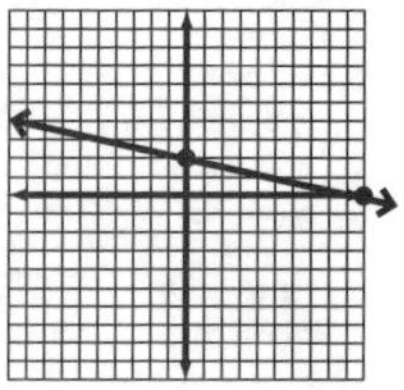

Name______________________ (8.EE.B.6, 8.F.A.1)

Graphing Linear Equations

Another way to graph a linear equation is to use the **slope-intercept form**. A linear equation written in slope-intercept form is $y = mx + b$, where m is the slope of the line and b is the y-intercept.

Example: $y = 3x + 2$, $m = 3$, y-int. $= 2$; $y = {}^{-}5x + 1$, $m = {}^{-}5$, y-int. $= 1$; $y = x - 7$, $m = 1$, y-int. $= {}^{-}7$

You can rewrite an equation in slope-intercept form by adding or subtracting terms from both sides and simplifying.

To graph the equation, plot the y-intercept. Starting at the y-intercept, move up or down and then right or left according to the slope and make a new point.

Example: If $m = 4$ and $b = 3$, the y-intercept is (0, 3). The next point to graph would be (1, 7) since the slope is 4, or $\frac{4}{1}$.

Solve for y, state the m and b, and graph.

1. $y = {}^{-}3x + 5$
m = −3
b = 5

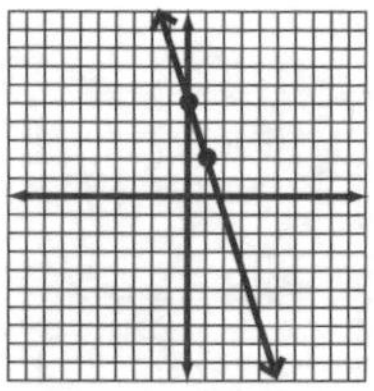

2. ${}^{-}8x + y = {}^{-}4$
m = 8
b = −4

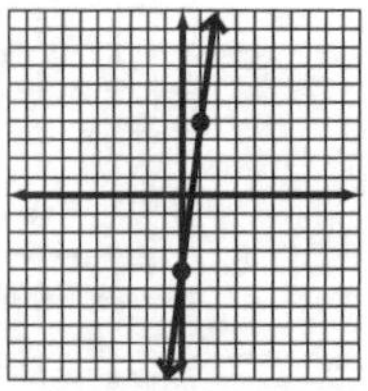

3. $x - y = 7$
m = 1
b = −7

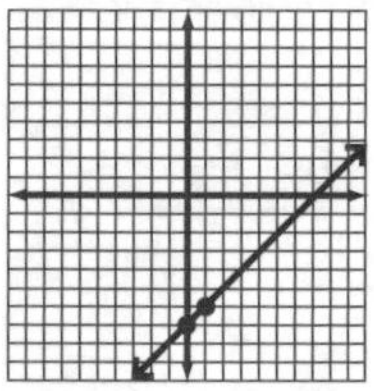

4. $2y = 3x - 4$
$\mathbf{m = \frac{3}{2}}$
b = −2

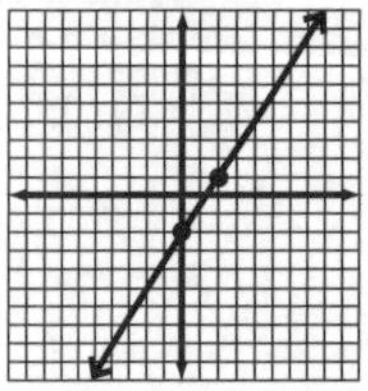

Name______________________ (8.EE.B.6, 8.F.A.1)

Graphing Linear Equations

Graph each equation using either slope-intercept form or x- and y-intercepts.

1. $2x + 3y = 9$

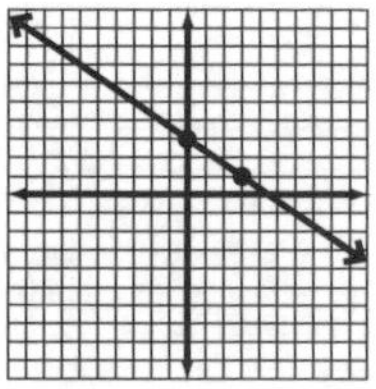

2. $x - 3y + 9 = 0$

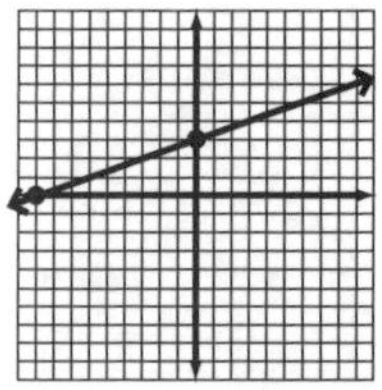

3. $6x + 2y = 12$

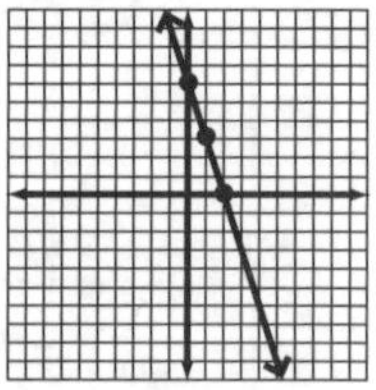

4. $y = 2x - 4$

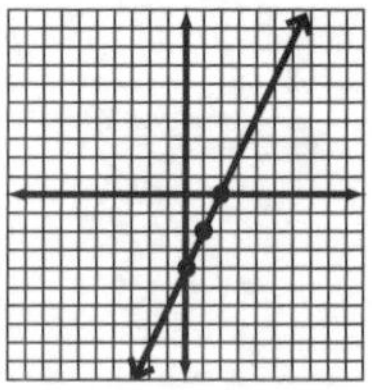

5. $5x - y = 7$

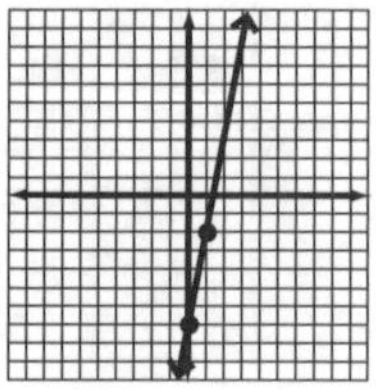

6. $x - 3y + 6 = 0$

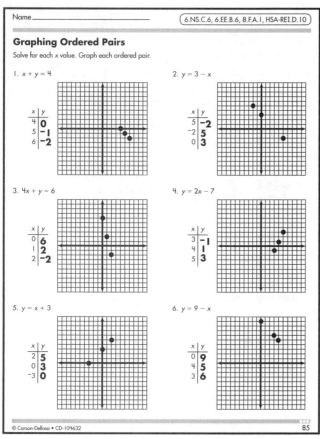

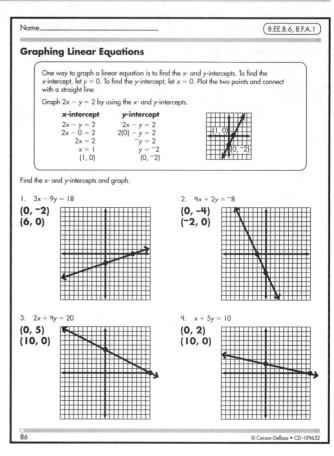

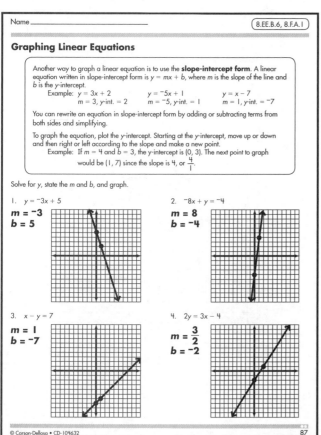

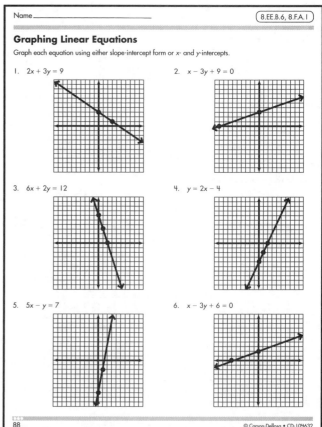

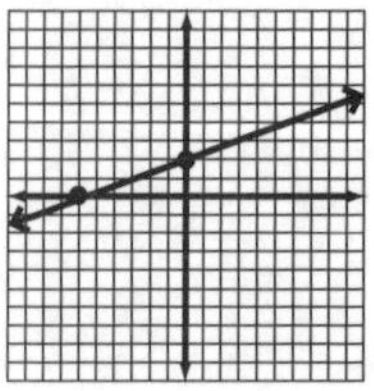

Answer Key

Name ____________________ (8.EE.B.6, 8.F.A.3)

Writing Linear Equations

This equation, $y = mx + b$, is the **slope-intercept of a straight line.** For all equations of the form $y = mx + b$, m is the slope of the line. The y-intercept is $(0, b)$.

Find the equation of a line using the slope-intercept form: $y = mx + b$.

1. $m = 4$, y-int. $= {}^-1$
   $\mathbf{y = 4x - 1}$
2. $m = {}^-2$, y-int. $= {}^-5$
   $\mathbf{y = {}^-2x - 5}$
3. $m = 1$, y-int. $= 2$
   $\mathbf{y = x + 2}$
4. $m = \frac{7}{5}$ $b = {}^-2$
   $\mathbf{y = \frac{7}{5}x - 2}$
5. $m = \frac{3}{4}$ $b = \frac{2}{3}$
   $\mathbf{y = \frac{3}{4}x + \frac{2}{3}}$
6. $m = \frac{3}{7}$ $b = \frac{1}{3}$
   $\mathbf{y = \frac{3}{7}x + \frac{1}{3}}$
7. $m = \frac{3}{5}$ $b = \frac{1}{5}$
   $\mathbf{y = \frac{3}{5}x + \frac{1}{5}}$
8. $m = {}^-4$ $b = \frac{3}{4}$
   $\mathbf{y = {}^-4x + \frac{3}{4}}$
9. $m = 0$ $b = 5$
   $\mathbf{y = 5}$
10. $m = \frac{1}{4}$ $b = \frac{2}{3}$
   $\mathbf{y = \frac{1}{4}x + \frac{2}{3}}$

Name ____________________ (8.EE.B.6, 8.F.A.3)

Writing Linear Equations

Slope-Intercept Formula
$y = mx + b$
m: slope
b: y-intercept containing coordinate points (o, b)

$m = 4$, passing through points $(1, 2)$
Using this information, $\{m = 4, x = 1, y = 2\}$.
Substitute into $y = mx + b$ to find b.
$2 = (4)(1) + b \longrightarrow 2 = 4 + b$
$\longrightarrow 2 - 4 = 4 - 4 + b \longrightarrow {}^-2 = b$
With the information $m = 4$, $b - {}^-2$, write the equation as $y = 4x - 2$.

Find the equation of the line with the given slope passing through the indicated point (P).

1. $p = ({}^-2, {}^-6)$, $m = 3$
   $b =$ **0**
   Equation **$y = 3x$**
2. $p = ({}^-4, 3)$, $m = 1$
   $b =$ **7**
   Equation **$y = x + 7$**
3. $p = (3, 5)$, $m = 0$
   $b =$ **5**
   Equation **$y = 5$**
4. $p = (5, 7)$, $m = 3$
   $b =$ **$^-8$**
   Equation **$y = 3x - 8$**
5. $p = ({}^-7, {}^-7)$, $m = {}^-7$
   $b =$ **$^-56$**
   Equation **$y = {}^-7x - 56$**
6. $p = (2, {}^-6)$, $m = 4$
   $b =$ **$^-14$**
   Equation **$y = 4x - 14$**
7. $p = (2, 4)$, $m = 4$
   $b =$ **$^-4$**
   Equation **$y = 4x - 4$**
8. $p = (6, {}^-1)$, $m = {}^-5$
   $b =$ **29**
   Equation **$y = {}^-5x + 29$**
9. $p = ({}^-1, 1)$, $m = 2$
   $b =$ **3**
   Equation **$y = 2x + 3$**
10. $p = ({}^-1, {}^-6)$, $m = 2$
   $b =$ **$^-4$**
   Equation **$y = 2x - 4$**
11. $p = (4, 5)$, $m = {}^-2$
   $b =$ **13**
   Equation **$y = {}^-2x + 13$**
12. $p = (2, 6)$, $m = {}^-6$
   $b =$ **18**
   Equation **$y = {}^-6x + 18$**

Name ____________________ (8.EE.B.6, 8.F.A.3)

Writing Linear Equations

Write the equation of the line that passes through the point and has the given slope. Put each equation in slope-intercept form.

1. $p = (2, 3)$, $m = 4$
   $\mathbf{y = 4x - 5}$
2. $p = ({}^-5, {}^-4)$, $m = {}^-8$
   $\mathbf{y = {}^-8x - 44}$
3. $p = ({}^-5, 7)$, $m = {}^-3$
   $\mathbf{y = {}^-3x - 8}$
4. $p = (2, 4)$, $m = {}^-1$
   $\mathbf{y = {}^-x + 6}$
5. $p = ({}^-1, 2)$, $m = 1$
   $\mathbf{y = x + 3}$
6. $p = (6, {}^-8)$, $m = 2$
   $\mathbf{y = 2x - 20}$
7. $p = (1, {}^-1)$, $m = \frac{1}{2}$
   $\mathbf{y = \frac{1}{2}x - \frac{3}{2}}$
8. $p = ({}^-4, {}^-1)$, $m = -\frac{1}{2}$
   $\mathbf{y = -\frac{1}{2}x - 3}$
9. $p = (6, 9)$ $m = 5$
   $\mathbf{y = 5x - 21}$
10. $p = ({}^-1, 2)$ $m = {}^-1$
   $\mathbf{y = {}^-x + 1}$
11. $p = ({}^-2, 9)$ $m = 3$
   $\mathbf{y = 3x + 15}$
12. $p = ({}^-2, {}^-1)$ $m = 2$
   $\mathbf{y = 2x + 3}$
13. $p = (3, 10)$ $m = {}^-5$
   $\mathbf{y = {}^-5x + 25}$
14. $p = (7, 8)$ $m = 4$
   $\mathbf{y = 4x - 20}$

Name ____________________ (HSA-REI.D.12)

Graphing Linear Inequalities

Graph the line $y > 2x + 3$.

1. $m = \frac{2}{1}$ $b = 3$
2. If $>$ or $<$, connect the points with a dotted line. 
3. If $\geq$ or $\leq$, connect the points with a solid line. 
   The coordinate plane is now divided into 2 regions.
4. Test any (x, y) on each side of the line in the original inequality.

$y > 2x + 3$
Test point A $({}^-1, 4)$.
Is $4 > 2({}^-1) + 3$?
$4 > {}^-2 + 3$
$4 > 1 \longrightarrow$ true
(Shade this region)
Test point A.

$y > 2x + 3$
Test point B $(0, 0)$.
Is $0 > (0) + 3$?
$0 > 0 + 3$
$0 > 3 \longrightarrow$ false
(Do not shade this region)
Test point B.

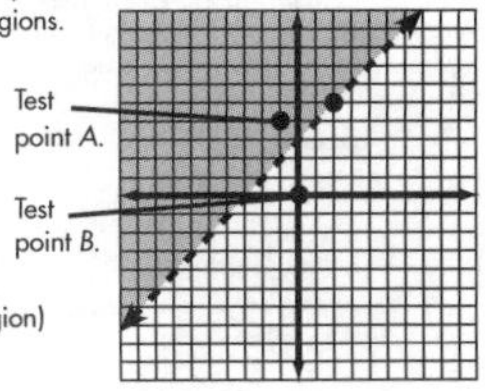

Graph the solution set.

1. $x + 4y > 8$ $\mathbf{y > -\frac{1}{4}x + 2}$

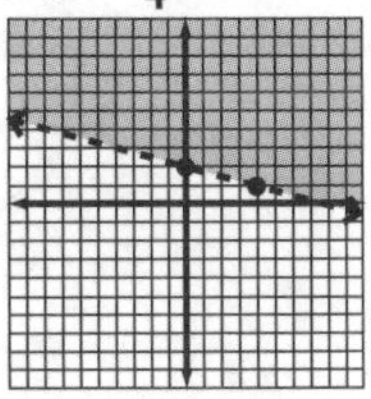

2. ${}^-2x + 2y \geq 10$ $\mathbf{y \geq x + 5}$

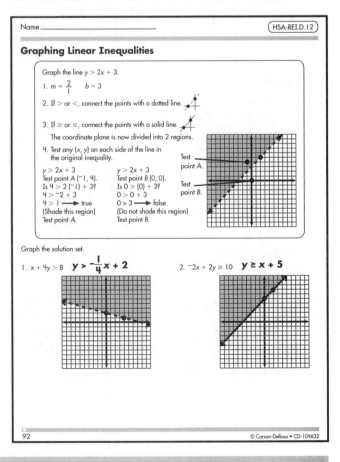

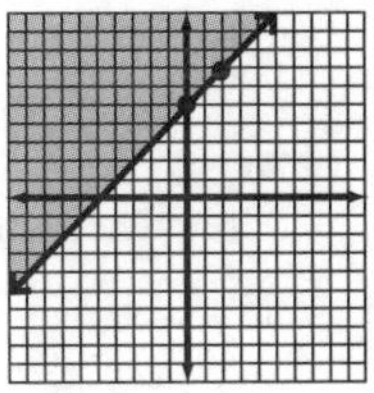

Answer Key

Name____________________ HSA-REI.D.12

Graphing Linear Inequalities

Graph the solution set.

1. $3x + 2y \geq 6$ **$y \geq -\frac{3}{2}x + 3$**

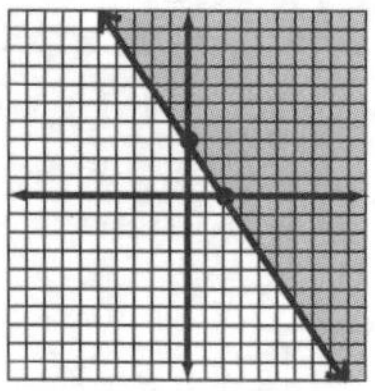

2. $2x + y < 4$ **$y < -2x + 4$**

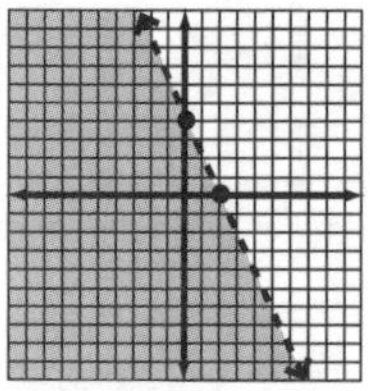

3. $6x - 3y > 15$ **$y < 2x - 5$**

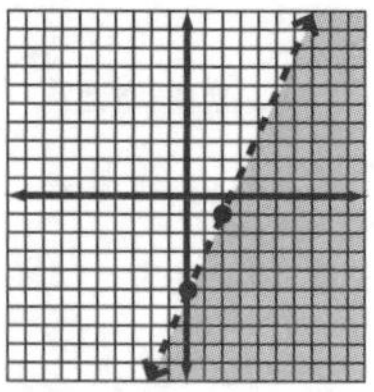

4. $2x + 3y \geq 6$ **$y \geq -\frac{2}{3}x + 2$**

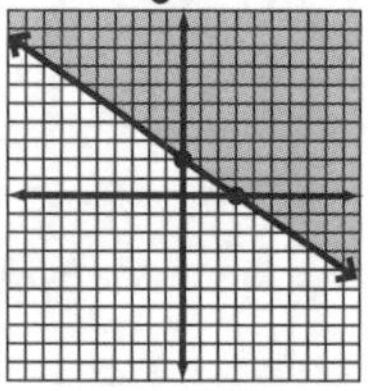

5. $3x - 4y \leq 12$ **$y \geq \frac{3}{4}x - 3$**

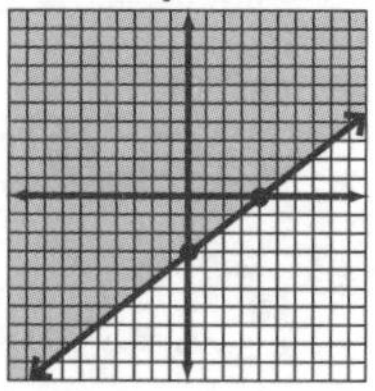

6. $4x + 2y < 6$ **$y < -2x + 3$**

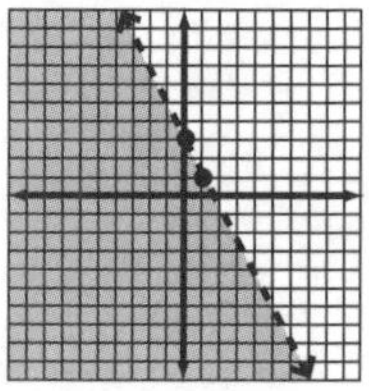

Name____________________ HSA-REI.D.12

Graphing Linear Inequalities

Graph the solution set.

1. $x + 2y < 0$ **$y < -\frac{x}{2}$**

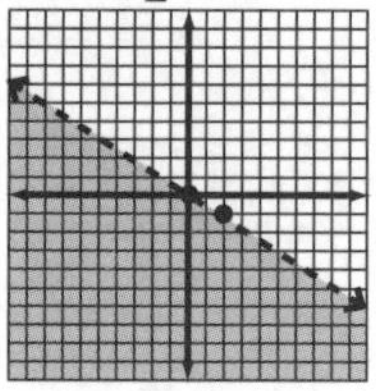

2. $5x - 2y \leq 10$ **$y \geq \frac{5}{2}x - 5$**

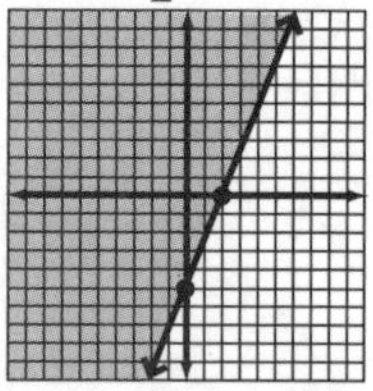

3. $6x - 3y > 18$ **$y < 2x - 6$**

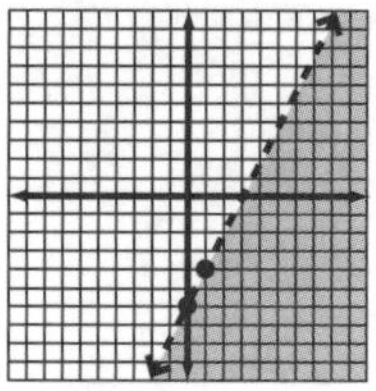

4. $2x - 5y < 10$ **$y > \frac{2}{5}x - 2$**

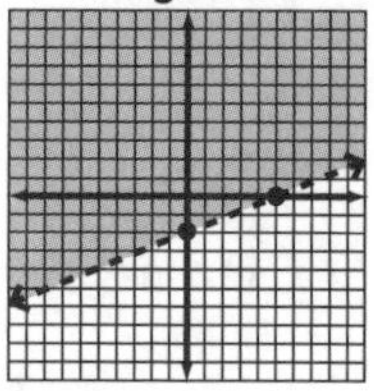

5. $-5x + 5y \leq 15$ **$y \leq x + 3$**

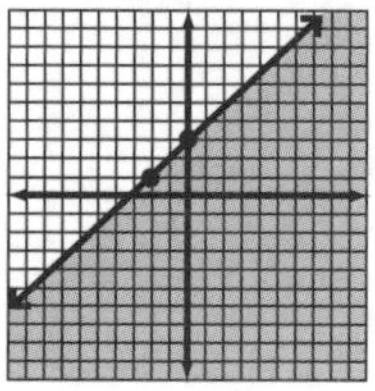

6. $y + 6 > 0$ **$y > -6$**

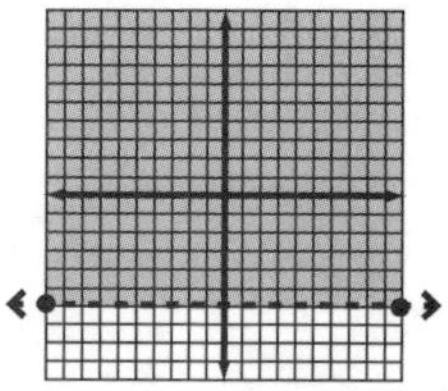

Name____________________ 8.EE.C.8b, HSA-REI.C.6, HSA-REI.D.11

Solving Systems of Equations by Graphing

When two or more equations are considered together, it is called a **system of equations**. The following example is a system of two linear equations in two variables.

$x + 2y = 4$
$2x + y = -1$

The graphs of these equations are straight lines.

An ordered pair that is a solution of each equation of the system is a solution of the system of equations in two variables. The solution of a system of linear equations can be found by graphing the lines of the system. The solution of the system of equations is the point where the lines intersect.

Solve by graphing:
$x + 2y = 4$
$2x + y = -1$
Graph each line and find the point of intersection.
The solution is $(-2, 3)$ because the ordered pair lies on each line.

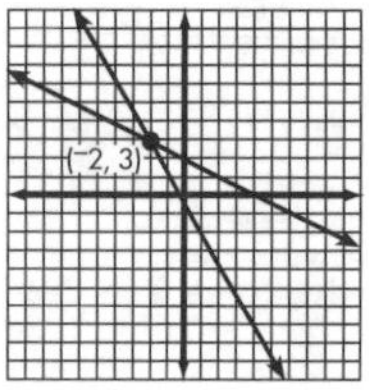

Solve by graphing.

1. $x + y = 7$
$3x - y = -3$
(1, 6)

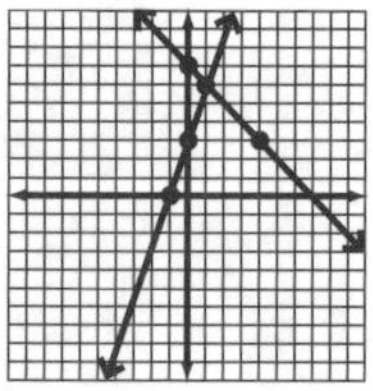

2. $x + y = 4$
$x - y = 6$
(5, −1)

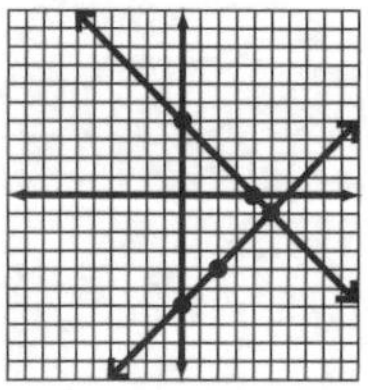

Name____________________ 8.EE.C.8b, HSA-REI.C.6, HSA-REI.D.11

Solving Systems of Equations by Graphing

Solve by graphing.

1. $3x - 4y = 12$
$2x + 4y = -12$
(0, −3)

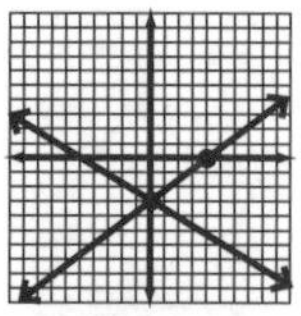

2. $4x - 2y = 8$
$y = 2$
(3, 2)

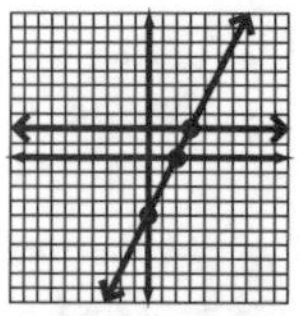

3. $x - y = 1$
$x + 2y = 10$
(4, 3)

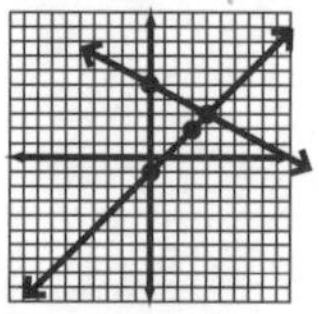

4. $x = 5$
$y = -1$
(5, −1)

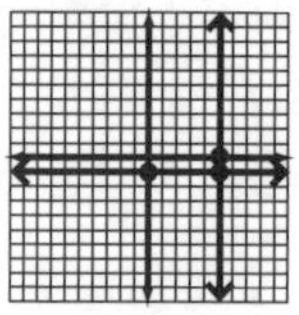

5. $y = 2x + 4$
$y = x + 6$
(2, 8)

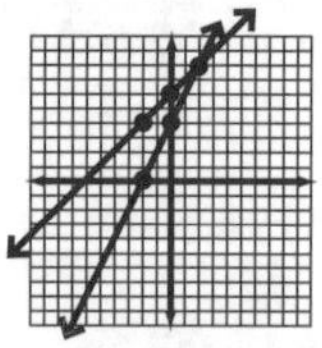

6. $x - y = -4$
$3x - y = -12$
(−4, 0)

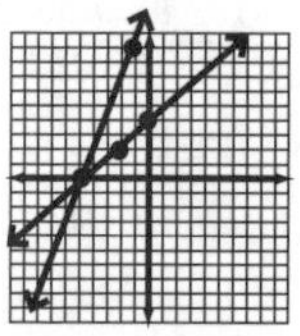

7. $x + y = 3$
$x - y = 5$
(4, −1)

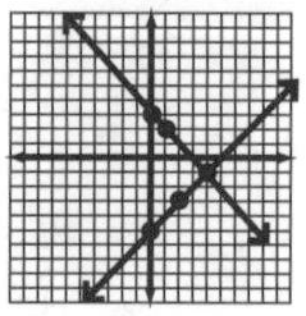

8. $x = 4$
$6x - 2y = 4$
(4, 10)

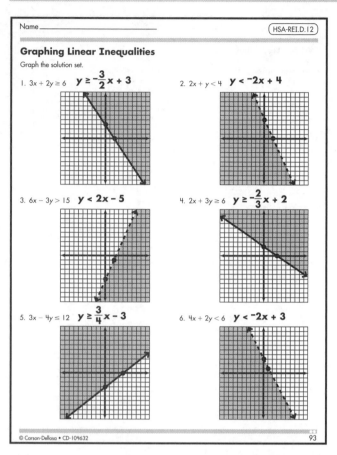

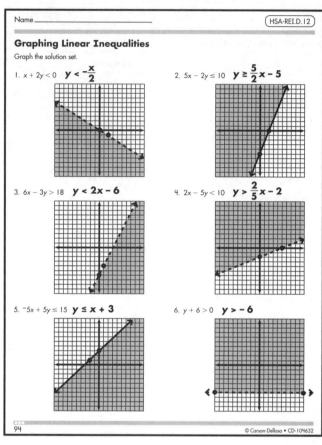

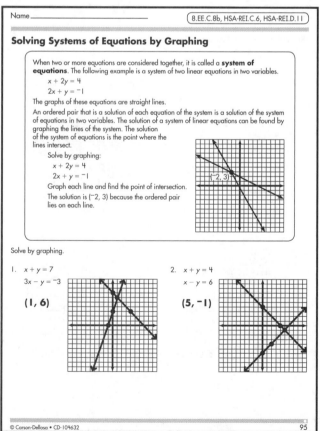

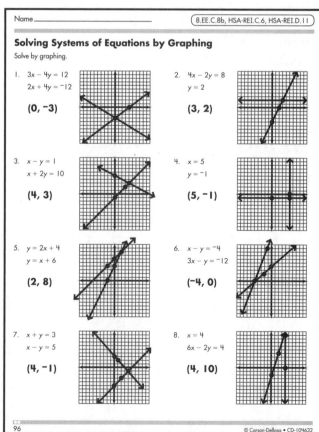

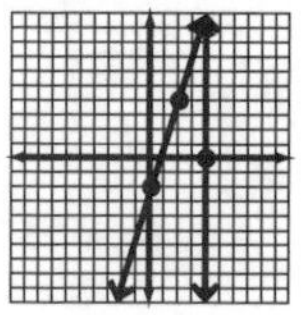

Answer Key

Name____________________ (8.EE.C.8b, HSA-REI.C.6, HSA-REI.D.11)

Solving Systems of Equations by Graphing

Solve by graphing.

1. $y = x - 3$
   $x = 2y + 6$
   (0, −3)

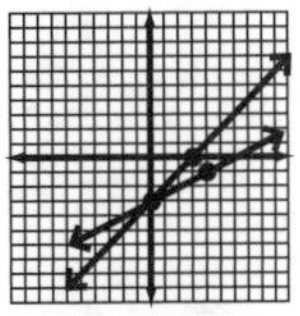

2. $-2x + y = 0$
   $6x + y = -16$
   (−2, −4)

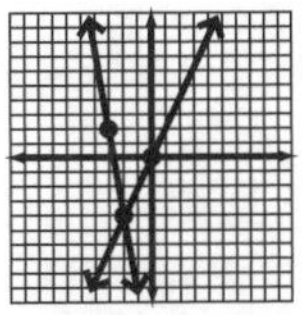

3. $y = 3x + 5$
   $y = 1 - x$
   (−1, 2)

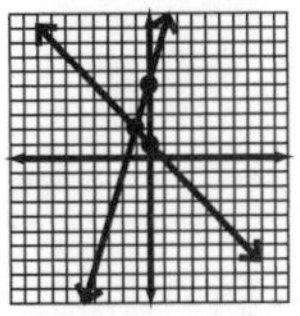

4. $y = x - 5$
   $y + 2x = 1$
   (2, −3)

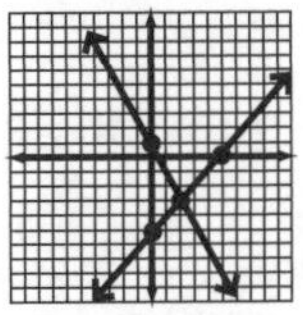

5. $x - 4y = 2$
   $x + 2y = 8$
   (6, 1)

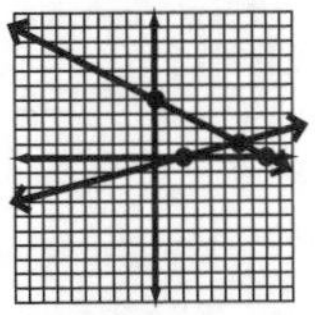

6. $x + y = 6$
   $3x - y = 2$
   (2, 4)

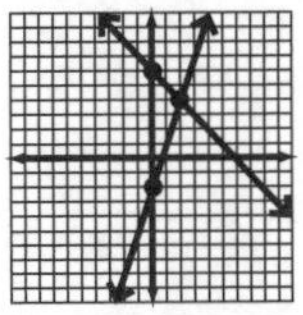

7. $2x + y = 6$
   $3x + y = 12$
   (6, −6)

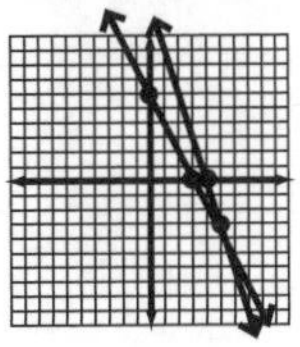

8. $y = x + 3$
   $2y = 3x + 1$
   (5, 8)

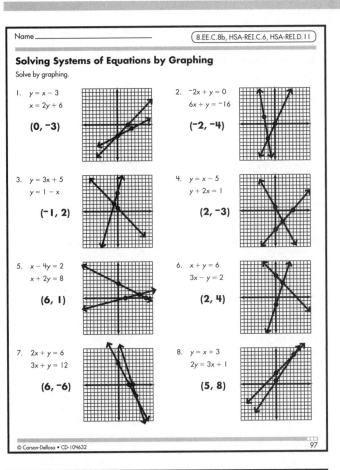

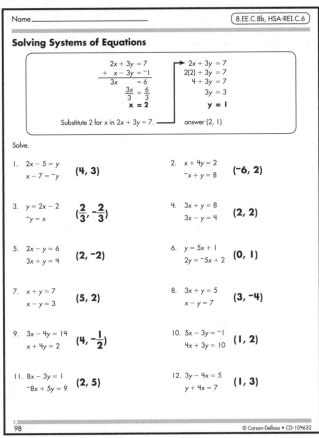

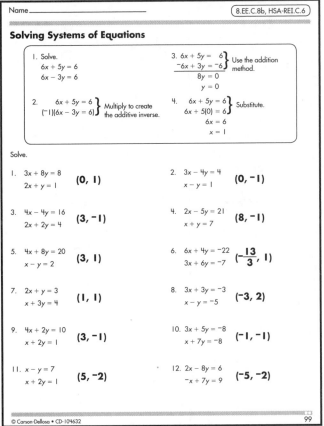

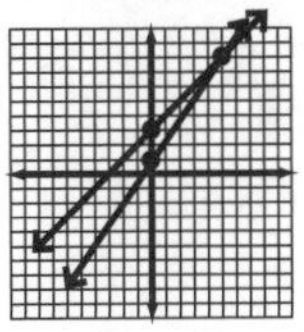

Name____________________ (8.EE.C.8b, HSA-REI.C.6)

Solving Systems of Equations

$$\begin{array}{r} 2x + 3y = 7 \\ +\ \ x - 3y = -1 \\ \hline 3x \quad\quad = 6 \\ \frac{3x}{3} = \frac{6}{3} \\ \mathbf{x = 2} \end{array} \qquad \begin{array}{r} 2x + 3y = 7 \\ 2(2) + 3y = 7 \\ 4 + 3y = 7 \\ 3y = 3 \\ \mathbf{y = 1} \end{array}$$

Substitute 2 for x in $2x + 3y = 7$. answer (2, 1)

Solve.

1. $2x - 5 = y$, $x - 7 = -y$ **(4, 3)**
2. $x + 4y = 2$, $-x + y = 8$ **(−6, 2)**
3. $y = 2x - 2$, $-y = x$ **($\frac{2}{3}$, $-\frac{2}{3}$)**
4. $3x + y = 8$, $3x - y = 4$ **(2, 2)**
5. $2x - y = 6$, $3x + y = 4$ **(2, −2)**
6. $y = 5x + 1$, $2y = -5x + 2$ **(0, 1)**
7. $x + y = 7$, $x - y = 3$ **(5, 2)**
8. $3x + y = 5$, $x - y = 7$ **(3, −4)**
9. $3x - 4y = 14$, $x + 4y = 2$ **(4, $-\frac{1}{2}$)**
10. $5x - 3y = -1$, $4x + 3y = 10$ **(1, 2)**
11. $8x - 3y = 1$, $-8x + 5y = 9$ **(2, 5)**
12. $3y - 4x = 5$, $y + 4x = 7$ **(1, 3)**

Name____________________ (8.EE.C.8b, HSA-REI.C.6)

Solving Systems of Equations

1. Solve.
   $6x + 5y = 6$
   $6x - 3y = 6$
2. $6x + 5y = 6$, $(-1)(6x - 3y = 6)$ } Multiply to create the additive inverse.
3. $6x + 5y = 6$, $-6x + 3y = -6$ } Use the addition method.
   $8y = 0$
   $y = 0$
4. $6x + 5y = 6$, $6x + 5(0) = 6$ } Substitute.
   $6x = 6$
   $x = 1$

Solve.

1. $3x + 8y = 8$, $2x + y = 1$ **(0, 1)**
2. $3x - 4y = 4$, $x - y = 1$ **(0, −1)**
3. $4x - 4y = 16$, $2x + 2y = 4$ **(3, −1)**
4. $2x - 5y = 21$, $x + y = 7$ **(8, −1)**
5. $4x + 8y = 20$, $x - y = 2$ **(3, 1)**
6. $6x + 4y = -22$, $3x + 6y = -7$ **($-\frac{13}{3}$, 1)**
7. $2x + y = 3$, $x + 3y = 4$ **(1, 1)**
8. $3x + 3y = -3$, $x - y = -5$ **(−3, 2)**
9. $4x + 2y = 10$, $x + 2y = 1$ **(3, −1)**
10. $3x + 5y = -8$, $x + 7y = -8$ **(−1, −1)**
11. $x - y = 7$, $x + 2y = 1$ **(5, −2)**
12. $2x - 8y = 6$, $-x + 7y = 9$ **(−5, −2)**

Name____________________ (8.EE.C.8b, HSA-REI.C.6)

Solving Systems of Equations

Solve.

1. $y = 3 - 2x$, $y = 2 - 3x$ **(−1, 5)**
2. $x + y = 5$, $x = y + 7$ **(6, −1)**
3. $x - y = 1$, $2x + y = 8$ **(3, 2)**
4. $3x - y = 9$, $y = x + 5$ **(7, 12)**
5. $3x + 4y = 26$, $-2x + y = 1$ **(2, 5)**
6. $y = 2x + 3$, $y = 4x + 4$ **($-\frac{1}{2}$, 2)**
7. $2x + 7y = 8$, $x + 5y = 7$ **(−3, 2)**
8. $y = 4x + 4$, $y = 2x + 8$ **(2, 12)**
9. $x + 3y = 17$, $2x + 3y = 22$ **(5, 4)**
10. $4x - 7y = 9$, $y = x - 3$ **(4, 1)**
11. $8x - 5y = 9$, $y = 2x - 4$ **($\frac{11}{2}$, 7)**
12. $2x + 4y = -2$, $3x + y = 7$ **(3, −2)**
13. $3x + y = 5$, $2x + 3y = 8$ **(1, 2)**
14. $2x + 6y = 24$, $x - 4y = -2$ **(6, 2)**

Answer Key

Name____________________ HSN-RN.A.2

Operations With Radicals

Simplify: $\sqrt{49x^4y^8} = 7\sqrt{x^4y^8} = 7x^2y^4$

Multiply: $\sqrt{3a} \bullet \sqrt{4a} = \sqrt{12a^2} = \sqrt{3} \bullet 4 \bullet a^2 = 2a\sqrt{3}$

Divide: $\sqrt{\frac{18}{2}} = \sqrt{9} = 3$ $\qquad \sqrt{\frac{9}{25}} = \frac{\sqrt{9}}{\sqrt{25}} = \frac{3}{5}$

Add or subtract: $2\sqrt{y} + 3\sqrt{y} + \sqrt{y} = 6\sqrt{y}$ $\qquad \sqrt{4x} + 3\sqrt{x} = 2\sqrt{x} + 3\sqrt{x} = 3\sqrt{x}$

Simplify.

1. $\sqrt{x^2y^{10}} =$ $\mathbf{xy^5}$
2. $\sqrt{27x^8} =$ $\mathbf{3x^4\sqrt{3}}$
3. $\sqrt{x^{16}} =$ $\mathbf{x^8}$
4. $\sqrt{125b^{15}} =$ $\mathbf{3b^7\sqrt{5b}}$
5. $\sqrt{x^2y^4} \bullet 2\sqrt{xy} =$ $\mathbf{2xy^2\sqrt{xy}}$
6. $\sqrt{9} \bullet \sqrt{32} =$ $\mathbf{12\sqrt{2}}$
7. $3\sqrt{5} \bullet 2\sqrt{4} =$ $\mathbf{12\sqrt{5}}$
8. $2\sqrt{4x^3y} \bullet y\sqrt{x^5y^7} =$ $\mathbf{4x^4y^5}$
9. $\sqrt{\frac{36}{9}} =$ $\mathbf{2}$
10. $\sqrt{\frac{27x}{3x}} =$ $\mathbf{3}$
11. $3\sqrt{x^3} - 4\sqrt{x^3} =$ $\mathbf{-x\sqrt{x}}$
12. $3\sqrt{2y} + 2\sqrt{2y} =$ $\mathbf{5\sqrt{2y}}$
13. $2\sqrt{y} - 4\sqrt{y} =$ $\mathbf{-2\sqrt{y}}$
14. $3y\sqrt{2y} - y\sqrt{2y} =$ $\mathbf{2y\sqrt{2y}}$

Name____________________ HSN-RN.A.2

Operations With Radicals

Simplify.

1. $\sqrt{x^{14}y^6} =$ $\mathbf{x^7y^3}$
2. $\sqrt{169y^{12}} =$ $\mathbf{13y^6}$
3. $\sqrt{16x^4} =$ $\mathbf{4x^2}$
4. $\sqrt{8x^3} =$ $\mathbf{2x\sqrt{2x}}$
5. $\sqrt{81x^6} =$ $\mathbf{9x^3}$
6. $\sqrt{25x^6} =$ $\mathbf{5x^3}$
7. $4\sqrt{9x^3} \bullet 3\sqrt{4x} =$ $\mathbf{72x^2}$
8. $x\sqrt{5x^3y} \bullet x\sqrt{5x^2y} =$ $\mathbf{5x^4y\sqrt{x}}$
9. $2\sqrt{9x^2} \bullet 2\sqrt{4x^2} =$ $\mathbf{24x^2}$
10. $6\sqrt{9xy} \bullet 4\sqrt{2xy} =$ $\mathbf{72xy\sqrt{2}}$
11. $2\sqrt{4x^3y} \bullet 3\sqrt{3a^2b^2} =$ $\mathbf{12abx\sqrt{3xy}}$
12. $\sqrt{\frac{x^2}{25}} =$ $\mathbf{\frac{x}{5}}$
13. $\sqrt{\frac{8x^3}{2x}} =$ $\mathbf{2x}$
14. $\sqrt{\frac{18x^3}{2x}} =$ $\mathbf{3x}$
15. $x\sqrt{27} + x\sqrt{12} =$ $\mathbf{5x\sqrt{3}}$
16. $3\sqrt{6x} + 5\sqrt{6x} =$ $\mathbf{8\sqrt{6x}}$
17. $4\sqrt{2x^3} + 3\sqrt{2x^3} =$ $\mathbf{7x\sqrt{2x}}$
18. $2\sqrt{50} - 4\sqrt{8} - 3\sqrt{72} =$ $\mathbf{-16\sqrt{2}}$

Name____________________ HSN-RN.A.2

Operations with Radicals

Simplify.

1. $\sqrt{x^9y^9} =$ $\mathbf{x^4y^4\sqrt{xy}}$
2. $\sqrt{54x^8} =$ $\mathbf{3x^4\sqrt{6}}$
3. $\sqrt{9a^4b^8} =$ $\mathbf{3a^2b^4}$
4. $\sqrt{49x^4y^2} =$ $\mathbf{7x^2y}$
5. $5\sqrt{2x^6y} \bullet 3\sqrt{3x^3y^5} =$ $\mathbf{15x^4y^3\sqrt{6x}}$
6. $4\sqrt{8a^6b} \bullet 4\sqrt{8a^4b^4} =$ $\mathbf{128a^5b^2\sqrt{b}}$
7. $3\sqrt{2x^3} \bullet 3\sqrt{3x^2y^2} =$ $\mathbf{9x^2y\sqrt{6x}}$
8. $5\sqrt{4a} \bullet 2\sqrt{6a} =$ $\mathbf{20a\sqrt{6}}$
9. $x\sqrt{3x} \bullet \sqrt{3x^3} =$ $\mathbf{3x^3}$
10. $\sqrt{2x^4} \bullet \sqrt{10x^2y^2} =$ $\mathbf{2x^3y\sqrt{5}}$
11. $\sqrt{\frac{9}{64}} =$ $\mathbf{\frac{3}{8}}$
12. $\sqrt{\frac{50x^2}{2}} =$ $\mathbf{5x}$
13. $\sqrt{\frac{49x^2}{25x^3}} =$ $\mathbf{\frac{7}{5\sqrt{x}}}$
14. $\sqrt{\frac{12x^2}{60}} =$ $\mathbf{\frac{x}{\sqrt{5}}}$
15. $\sqrt{\frac{3x^7}{108y^2}} =$ $\mathbf{\frac{x^3\sqrt{x}}{6y}}$
16. $4\sqrt{y^3} - 2\sqrt{y^3} =$ $\mathbf{2y\sqrt{y}}$
17. $y\sqrt{y^4} - 2y\sqrt{y^4} =$ $\mathbf{-y^3}$
18. $4x\sqrt{x^3} + 2x\sqrt{x^3} =$ $\mathbf{6x^2\sqrt{x}}$
19. $4\sqrt{x} - 2\sqrt{x} - 3\sqrt{x} + 5\sqrt{x} =$ $\mathbf{4\sqrt{x}}$
20. $3\sqrt{4x^2y} - 8y\sqrt{y} =$ $\mathbf{(6x - 8y)\sqrt{y}}$

Congratulations!

____

receives this award for

____

Signed ______________________________ **Date** ________________

$x + 6 = {}^{-}6$	$x - 3 = {}^{-}6$	$3x \div 3 = 12$	$\frac{x}{2} = 24$
$5x - 17 = 8$	$\frac{x}{3} = {}^{-}6$	${}^{-}7x + {}^{-}4 = 17$	$x - 4 = {}^{-}3$
$x + 5 = {}^{-}7$	$x - 4 = 6$	$x + 9 = {}^{-}12$	$4 - x = 6$
${}^{-}4x = 36$	$x + 4 = {}^{-}10$	$x - 5 = {}^{-}3$	$3x - 3 = 12$

$x = 48$	$x = 12$	$x = -3$	$x = -12$
$x = 1$	$x = -3$	$x = -18$	$x = 5$
$x = -2$	$x = -21$	$x = 10$	$x = -12$
$x = 5$	$x = 2$	$x = -14$	$x = -9$

$(3xy)^{-1}$ © CD	$(\frac{3}{4})^{-2}$ © CD	$(x^3y^2)^3$ © CD	$(\frac{x^3}{y^2})^4$ © CD
$(3x^2)(4x^3)$ © CD	$\frac{(x^6 \cdot x^4)}{x^2}$ © CD	$\frac{x^{12}}{x^3}$ © CD	$\frac{18x^3}{3x}$ © CD
$(2x^3)^3$ © CD	$(6x^{-2})^2$ © CD	$7x^{-6}$ © CD	$x^3 \cdot x \cdot x \cdot x^2$ © CD
$(3x^2y^2)^3$ © CD	$(-7x^3)^{-2}$ © CD	$(-3)^{-2}$ © CD	$x^4 \cdot x^4 \cdot x^4$ © CD

$\frac{x^{12}}{y^8}$	x^9y^6	$\frac{16}{9}$	$\frac{1}{3xy}$
$6x^2$	x^9	x^8	$12x^5$
x^7	$\frac{7}{x^6}$	$\frac{36}{x^4}$	$8x^9$
x^{12}	$\frac{1}{9}$	$\frac{1}{49x^6}$	$27x^6y^6$

2^{-3} © CD	$x^2 \cdot x^4$ © CD	$\frac{x^6}{x^3}$ © CD	$(-2)^3$ © CD
$5x^{-3}$ © CD	$(\frac{1}{2})^{-3}$ © CD	$(x^3)^3$ © CD	$x^3 \cdot x^2 \cdot x^3$ © CD
$(x^6)^3$ © CD	$x^3 \cdot x^3$ © CD	$(x^8)^2$ © CD	$(5x^3)^2$ © CD
$x^6 \cdot x^{-3}$ © CD	$3x^{-3}$ © CD	$(4x)^{-2}$ © CD	$x^3 \cdot x^{-7}$ © CD

$^{-}8$	x^3	x^6	$\frac{1}{8}$
x^8	x^9	8	$\frac{5}{x^3}$
$25x^6$	x^{16}	x^6	x^{18}
$\frac{1}{x^4}$	$\frac{1}{16x^2}$	$\frac{3}{x^3}$	x^3

Factor. $6x + 12y$ © CD	Factor. $3x + 4x^2$ © CD	Factor. $8 - 2y$ © CD	Factor. $36x^2 + 6x + 12$ © CD
Factor. $4x^2 + 64$ © CD	Factor. $x^2 - 4$ © CD	Factor. $9x^2 - 16$ © CD	Factor. $81 - x^2$ © CD
Factor. $x^2 - 9$ © CD	Factor. $36x^2 - 81$ © CD	Factor. $36x^2 - 49y^2$ © CD	Factor. $25x^2 - y^2$ © CD
Factor. $3x^2 - 36$ © CD	Factor. $5x^2 - 35$ © CD	Factor. $x^2 + 5x + 6$ © CD	Factor. $x^2 + 6x + 9$ © CD

$6(6x^2 + x + 2)$	$2(4 - y)$	$x(3 + 4x)$	$6(x + 2y)$
$(9 - x)(9 + x)$	$(3x - 4)(3x + 4)$	$(x - 2)(x + 2)$	$4(x^2 + 16)$
$(5x - y)(5x + y)$	$(6x + 7y)(6x - 7y)$	$9(2x - 3)(2x + 3)$	$(x - 3)(x + 3)$
$(x + 3)(x + 3)$	$(x + 3)(x + 2)$	$5(x^2 - 7)$	$3(x^2 - 12)$

Factor.

$x^2 + 9x + 14$

Factor.

$x^2 + 10x + 16$

Factor.

$x^2 - 8x + 15$

Factor.

$x^2 - 12x + 36$

Factor.

$x^2 - 10x + 24$

Factor.

$x^2 + 11x + 28$

Factor.

$x^2 - 5x - 6$

Factor.

$x^2 - x - 72$

Factor.

$x^2 - 3x - 10$

Factor.

$x^2 - 2x - 8$

Factor.

$x^2 + 4x - 21$

Factor.

$x^2 + 2x - 15$

Factor.

$x^2 + 2x - 24$

Factor.

$x^2 - 13x - 48$

Factor.

$3x^4 - 243$

Factor.

$2x^2 - 4x - 48$

$(x-6)(x-6)$	$(x-3)(x-5)$	$(x+8)(x+2)$	$(x+7)(x+2)$
$(x-9)(x+8)$	$(x-6)(x+1)$	$(x+7)(x+4)$	$(x-6)(x-4)$
$(x+5)(x-3)$	$(x+7)(x-3)$	$(x-4)(x+2)$	$(x-5)(x+2)$
$2(x-6)(x+4)$	$3(x+3)(x-3)$ (x^2+9)	$(x-16)(x+3)$	$(x+6)(x-4)$

$\sqrt{25}$ © CD	$\sqrt{9x^2}$ © CD	$\sqrt{144}$ © CD	$\sqrt{36}$ © CD
$\sqrt{81}$ © CD	$\sqrt{196x^4}$ © CD	$\sqrt{x^8}$ © CD	$\sqrt{x^6}$ © CD
$\sqrt{27}$ © CD	$\sqrt{20}$ © CD	$\sqrt{54}$ © CD	$\sqrt{x^{10}}$ © CD
$\sqrt{48}$ © CD	$\sqrt{4}$ © CD	$\sqrt{100}$ © CD	$\sqrt{121}$ © CD

± 6	± 12	$\pm 3x$	± 5
$\pm x^3$	$\pm x^4$	$\pm 14x^2$	± 9
$\pm x^5$	$\pm 3\sqrt{6}$	$\pm 2\sqrt{5}$	$\pm 3\sqrt{3}$
± 11	± 10	± 2	$\pm 4\sqrt{3}$